BestMasters

Mit **„BestMasters“** zeichnet Springer die besten Masterarbeiten aus, die an renommierten Hochschulen in Deutschland, Österreich und der Schweiz entstanden sind. Die mit Höchstnote ausgezeichneten Arbeiten wurden durch Gutachter zur Veröffentlichung empfohlen und behandeln aktuelle Themen aus unterschiedlichen Fachgebieten der Naturwissenschaften, Psychologie, Technik und Wirtschaftswissenschaften. Die Reihe wendet sich an Praktiker und Wissenschaftler gleichermaßen und soll insbesondere auch Nachwuchswissenschaftlern Orientierung geben.

Springer awards **“BestMasters”** to the best master’s theses which have been completed at renowned Universities in Germany, Austria, and Switzerland. The studies received highest marks and were recommended for publication by supervisors. They address current issues from various fields of research in natural sciences, psychology, technology, and economics. The series addresses practitioners as well as scientists and, in particular, offers guidance for early stage researchers.

More information about this series at http://www.springer.com/series/13198

Corina Keller

Chern-Simons Theory and Equivariant Factorization Algebras

Corina Keller
Institute of Mathematics
University of Zurich
Zürich, Switzerland

ISSN 2625-3577 ISSN 2625-3615 (electronic)
BestMasters
ISBN 978-3-658-25337-0 ISBN 978-3-658-25338-7 (eBook)
https://doi.org/10.1007/978-3-658-25338-7

Library of Congress Control Number: 2019931809

Springer Spektrum

This Springer Spektrum imprint is published by the registered company Springer Fachmedien Wiesbaden GmbH part of Springer Nature
The registered company address is: Abraham-Lincoln-Str. 46, 65189 Wiesbaden, Germany

Acknowledgement

I would like to acknowledge and thank Prof. Dr. Alberto Cattaneo of the Mathematical Institute at the University of Zurich for giving me the opportunity to write my master's thesis in his research group and for the helpful discussions and advice within the process.

I own particular gratitude to PD Dr. Alessandro Valentino, my supervisor, for providing support and guidance at all times during this thesis. I am extremely grateful for all the extended discussions, careful explanations and valuable suggestions that helped and encouraged me to improve my mathematical understanding. Your passion for mathematics is very inspiring.

I want to thank Prof. Dr. Thomas Gehrmann of the Physical Institute at the University of Zurich for taking the time to supervise this thesis and for the discussions about the physical aspects of the project.

I would also like to thank Dr. Angelika Schulz from Springer press for the friendly support and help concerning the publishing of this work.

Finally, I am extremely and deeply grateful to my parents and my sisters for their continuous support and encouragement. You are my biggest inspiration.

Corina Keller

Contents

1 Introduction

The notion of a *factorization algebra* was used by K. Costello and O. Gwilliams [CG16], [Gwi12] to describe the structure on the collection of observables in both classical- and quantum field theories. Motivated by their work, this master's thesis aims at studying the factorization algebra of *classical* observables arising from the perturbative facets of *abelian Chern-Simons theories*. For this purpose, we describe the local structure of the derived moduli space of flat abelian bundles over a closed oriented 3-manifold via its associated derived formal moduli problem. The mathematical structures that emerge on the observables of this classical perturbative field theory are then explored. In particular, we construct a homotopy action of the group of gauge transformations on the factorization algebra of observables, leading to the notion of an *equivariant* factorization algebra. This construction allows to go beyond a purely perturbative viewpoint and to encode topological features of the underlying abelian Lie group in the structure of the classical observables.

The study of the perturbative aspects of abelian Chern-Simons theories and its associated factorization algebras is an interplay of many different areas and concepts in mathematical physics. The chart displayed in figure 1.1, together with the following discussion, provides a first overview on the mathematical formalisms involved in this project and their manifestation in physics. In a first part of this introductory text, we want to provide some physical intuition for the use of derived deformation theory in perturbative classical field theory. In particular, we want to clarify the notion of classical observables in this context. In a second part, we explain why we use factorization algebras to characterize classical observables. We also show how we interpret physical observables in a gauge theory. In a last part, we give a brief introduction to Chern-Simons gauge theories. We show how the action for classical Chern-Simons field theories is constructed and describe the solutions to the corresponding Euler-Lagrange equations.

C. Keller, *Chern-Simons Theory and Equivariant Factorization Algebras*, BestMasters, https://doi.org/10.1007/978-3-658-25338-7_1

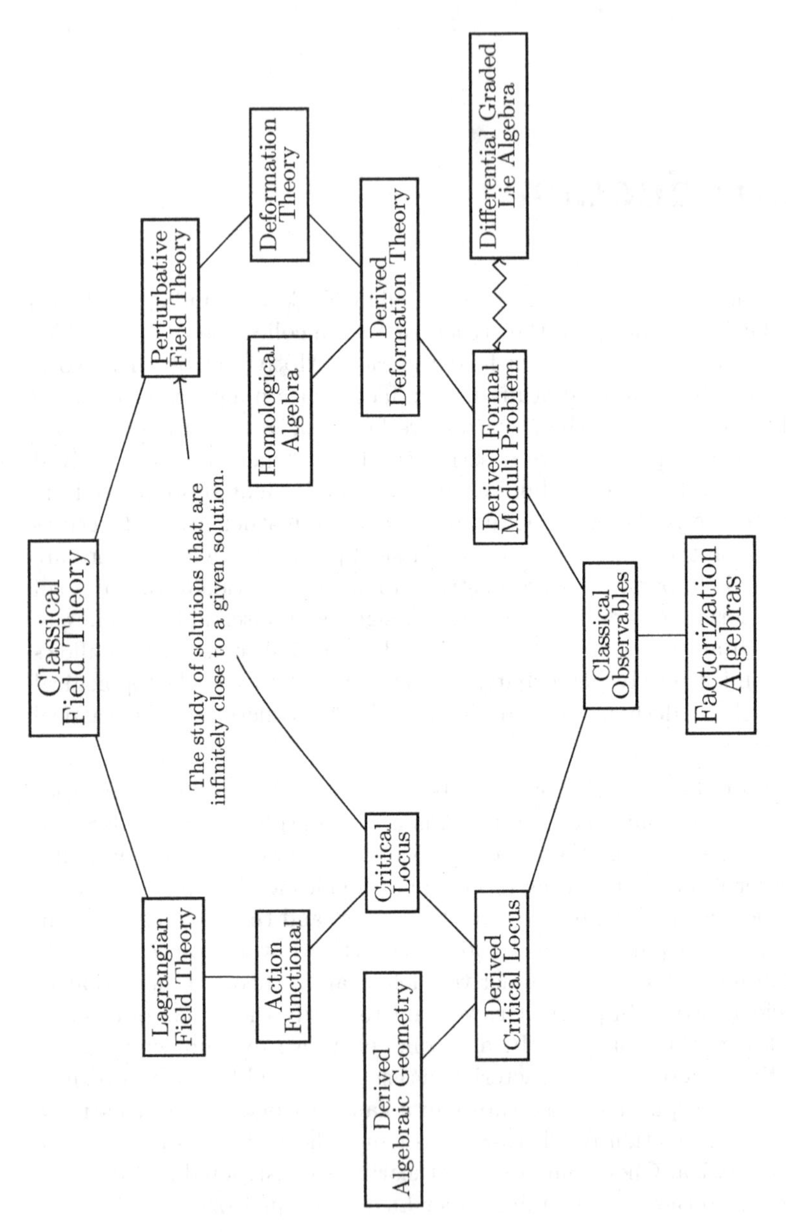

Figure 1.1: A sketch of the physical and mathematical areas explored in this thesis.

1.1 Classical Field Theory and Derived Deformations

The Lagrangian formulation of classical field theory is one of the most fundamental concepts of physics. The dynamical variables in the theory are the fields ϕ, which, in many situations, arise as sections in appropriate fiber bundles over spacetime. Lagrangian field theories are then described in term of a *Lagrangian density* $\mathcal{L}$, a local function of the fields and its derivatives

$$\mathcal{L} \equiv \mathcal{L}(\phi, \partial_\mu \phi).$$

The dynamics of the fields is governed by the *action functional* $\mathcal{S}$, which is given as the integral over spacetime of the Lagrangian density

$$\mathcal{S}[\phi] = \int d^4x \mathcal{L}(\phi, \partial_\mu \phi),$$

together with the *Euler-Lagrange equations* that arise as the critical points of the action. Hence, in Lagrangian field theories, the dynamical variables constitute the *space of fields* $\mathcal{M}$, which is often an infinite dimensional manifold, and the critical locus

$$\mathrm{Crit}(\mathcal{S}) = \{\phi \in \mathcal{M} \mid d\mathcal{S}(\phi) = 0\}$$

of the action functional $\mathcal{S} : \mathcal{M} \to \mathbb{R}$ is describing the classical physics of the system.

1.1.1 Derived Critical Locus

We want to sketch the essential ideas of describing the critical locus of an action functional in derived algebraic geometry. We refer to [CG16], [Gwi12] for a more extensive discussion.

Let us consider a finite dimensional smooth manifold X as a model for the space of fields and equip X with an action functional $\mathcal{S} : X \to \mathbb{R}$. Denote with $\mathrm{EL} \subset X$ the space of solutions to the Euler-Lagrange equations. Notice that we can view $d\mathcal{S}$ as a differential form on X. Hence, the ordinary critical locus is simply the fiber product of the graph of $d\mathcal{S}$ and the zero section inside the cotangent bundle T^*X. Pictorially, we have the following commutative diagram.

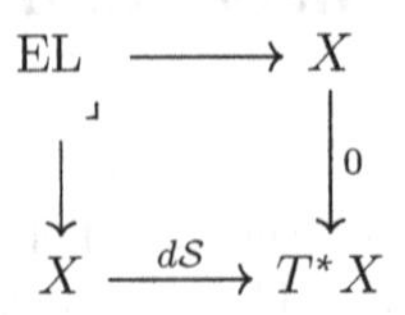

Algebraically, we can describe the commutative algebra $\mathcal{O}(\mathrm{EL})$ of functions on the critical locus EL as a tensor product

$$\mathcal{O}(\mathrm{EL}) = \mathcal{O}(\mathrm{graph}(d\mathcal{S})) \otimes_{\mathcal{O}(T^*X)} \mathcal{O}(X).$$

Notice that we write $\mathcal{O}(\mathrm{EL})$ to indicate any class of functions (smooth, polynomial, ...) considered on EL.

In many situations the critical locus, as described above, is not well-behaved. A possible way out is to consider the *derived critical locus* $\widetilde{\mathrm{EL}}$ of the action functional, which is defined to be the *derived intersection* of $\mathrm{graph}(d\mathcal{S}) \subset T^*X$, with the zero section $X \subset T^*X$. Vaguely, the idea here is that we don't force the vanishing of $d\mathcal{S}$ on the nose but only up to homotopy. More explicitly, in derived algebraic geometry functions on the derived intersection are defined by derived tensor products

$$\mathcal{O}(\widetilde{\mathrm{EL}}) = \mathcal{O}(\mathrm{graph}(d\mathcal{S})) \otimes^{\mathbb{L}}_{\mathcal{O}(T^*X)} \mathcal{O}(X).$$

To get a better understanding of the algebra of functions that determines the derived space of solutions, we have to consider an explicit model for the derived tensor product. Following [Vez11], the Koszul complex

$$(\mathrm{Sym}_{\mathcal{O}(X)}(TX[1]), d_{d\mathcal{S}}),$$

with differential $d_{d\mathcal{S}}$ given by contraction along $d\mathcal{S}$, is the *commutative differential graded (dg) algebra* of functions on $\widetilde{\mathrm{EL}}$. We explore the notion of (commutative) dg algebras in chapter 3.

Remark 1.1. Working with the derived critical locus is essentially what is know as the classical Batalin-Vilkovisky (BV) formalism. The basic idea for the use of the BV formalism in classical field theories is indicated in [CG16], [Gwi12]. For a formal treatment we refer to [Vez11].

1.1.2 Deformation Theory

In physics, we often study small perturbations of an exactly solvable problem, deforming a given solution by a small amount. In other words, we search for solutions to the Euler-Lagrange equation that are infinitely near a given solution ϕ_0. One way to think about this is to consider a series expansion around ϕ_0. More precisely, let ϵ be some small nilpotent parameter, i.e. $\epsilon^{n+1} = 0$ for some integer n, and consider the perturbation series

$$\phi \approx \phi_0 + \epsilon^1\phi_1 + \epsilon^2\phi_2 + \cdots + \epsilon^n\phi_n. \qquad (\diamond)$$

The Euler-Lagrange equation for ϕ simplifies to a system of differential equations classified by powers of ϵ. For $n \to \infty$ we recover ϕ as a formal power series in ϵ. With this procedure we describe the *formal neighborhood* of ϕ_0 in the space of solutions to the equations of motion.

Mathematically, these ideas are encoded in a formalism called *deformation theory.* The aspects of this theory used to express the perturbative viewpoint on classical field theory are at the focus of this thesis. In the first part of chapter 4 we thus spend some time to review classical formal deformations of mathematical structures. In particular, we give a functorial perspective on deformation theory in algebraic geometry, which allows a natural generalization to the framework of derived algebraic geometry.

Derived Formal Moduli Problems

The mathematical formalism of *derived* deformation theory combines ideas from homological algebra and classical deformation theory. For instance, the parameter ϵ in the perturbation series ($\diamond$) is in a derived setting possibly of non-zero cohomological degree.

Our interest in this thesis is in the local structure of the *derived moduli space* of classical solutions. A moduli space is a space parameterizing equivalence classes of mathematical structures, here, equivalence classes of solutions to the Euler-Lagrange equations. In chapter 4 we study the perturbative aspects of derived moduli spaces in terms of *derived formal moduli problems.* Roughly speaking, a formal moduli problem is a functor that captures the idea of describing the neighborhood of a given point in the moduli space of classical solutions.

Remark 1.2. There is a deep relation between dg Lie algebras and derived formal moduli problems. Namely, the Maurer-Cartan functor, sending certain dg Lie algebras to the simplicial set of their Maurer-Cartan elements, is a derived formal moduli problem [Get09]. The work of J. Lurie [Lur11] shows that *every* derived formal moduli problem is represented by a dg Lie algebra. We devote the second part of chapter 4 to talk about dg Lie algebras and their associated derived formal moduli problems.

1.1.3 Classical Observables

We want to assign to each region in spacetime a collection of classical observables. To that end, let M be a topological space, our spacetime, and let $U \subset M$ be an open subset. We denote $\mathrm{EL}(U)$ the space of solutions to the Euler-Lagrange equations on U. The *classical observables* $\mathrm{Obs}^{\mathrm{cl}}(U)$ arise as functions on the space of solutions

$$\mathrm{Obs}^{\mathrm{cl}}(U) = \mathcal{O}(\mathrm{EL}(U)) \in \mathbf{cAlg}_k.$$

In the derived world, we describe the local structure of the space of solutions by a derived formal moduli problem $\mathcal{EL}$, which is actually a sheaf on M. Hence, we can assign to every open subset the derived formal moduli problem $\mathcal{EL}(U)$ of solutions to the equations of motion on $U \subset M$. The collection of classical observables

$$\mathrm{Obs}^{\mathrm{cl}}(U) = \mathcal{O}(\mathcal{EL}(U)) \in \mathbf{cdgAlg}_k$$

is then the commutative dg algebra of functions on $\mathcal{EL}(U)$. In the next section we see that those classical observables carry a natural structure of a factorization algebra.

1.2 Factorization Algebras

Our goal is to motivate, in a physical language, the use of factorization algebras in field theories. For this purpose, we give a model problem of a freely moving particle that should make the idea of factorization algebras appear in a natural way. Our discussion follows [CG16], [Sto14].

1.2.1 Motivating Example: Free Particle in $\mathbb{R}^n$

Consider a free particle moving in some confined region of space, which for simplicity we assume to be all of $\mathbb{R}^n$. The realizable trajectories of the particle within some time interval $I \subset \mathbb{R}$ are given by the collection of maps $f : I \to \mathbb{R}^n$. These maps constitute the *space of fields*, denoted $\mathcal{M}(I)$, for any open subset $I \subset \mathbb{R}$. The classical observables of the theory correspond to the measurements we can perform on this system. In the following, we want to examine the structure and formal properties of those observables in the classical- and quantum setting.

Observables in Classical Mechanics

In classical field theories, the physical trajectories of the particle are constrained to satisfy the Euler-Lagrange equations arising as differential equations from the action functional. For our model problem, the action functional is given by the massless free field theory

$$S[\phi] = \int_{t \in I} \phi(t)\partial_t^2 \phi(t) dt,$$

where $\phi \in \mathcal{M}(I)$. We denote the space of maps $f : I \to \mathbb{R}^n$ that satisfy the equations of motion by

$$\mathrm{EL}(I) \subset \mathcal{M}(I).$$

The inclusion of open intervals $I \hookrightarrow J$ induces a linear map

$$r_I^J : \mathrm{EL}(J) \to \mathrm{EL}(I),$$

given by restriction. Following the discussion in the previous section, the *classical observables* $\mathrm{Obs}^{\mathrm{cl}}(I)$ are the functions on $\mathrm{EL}(I)$

$$\mathrm{Obs}^{\mathrm{cl}}(I) = \mathcal{O}(\mathrm{EL}(I)).$$

In other words, to each open subset $I \subset \mathbb{R}$ we assign the real vector space $\mathrm{Obs}^{\mathrm{cl}}(I)$ of classical observables on I. Notice that we obtain a linear *structure map*

$$m_J^I : \mathrm{Obs}^{\mathrm{cl}}(I) \to \mathrm{Obs}^{\mathrm{cl}}(J),$$

induced by the inclusion of open intervals $I \hookrightarrow J$. This linear structure map is compatible with composition, i.e. we have $m_K^I = m_K^J \circ m_J^I$ for open intervals $I \subset J \subset K$. Hence, this construction can be described by a functor

$$\mathrm{Obs}^{\mathrm{cl}} : \mathbf{Opens}(\mathbb{R}) \to \mathbf{Vect}_{\mathbb{R}},$$

where **Opens**($\mathbb{R}$) is the category of open intervals on $\mathbb{R}$ with morphisms given by inclusion. We say that the classical observables form a *precosheaf* on $\mathbb{R}$.

In the classical world we assume that the physical system is not disturbed by the process of conducting a measurement. Thus, we can combine measurements on a given time interval $I \subset \mathbb{R}$, that is, $\mathrm{Obs}^{\mathrm{cl}}(I)$ is the *commutative* algebra of functions on $\mathrm{EL}(I)$ with multiplication

$$\mathrm{Obs}^{\mathrm{cl}}(I) \otimes \mathrm{Obs}^{\mathrm{cl}}(I) \to \mathrm{Obs}^{\mathrm{cl}}(I).$$

Given the commutative structure on the classical observables, we see that for a finite collection of disjoint open time intervals $I_1, \ldots, I_k$ of an open subset $I \subset \mathbb{R}$ we have a *multiplication map*

$$m_J^{I_1,\ldots,I_k} : \mathrm{Obs}^{\mathrm{cl}}(I_1) \otimes \cdots \otimes \mathrm{Obs}^{\mathrm{cl}}(I_k) \to \mathrm{Obs}^{\mathrm{cl}}(I),$$

given by the composition

$$\mathrm{Obs}^{\mathrm{cl}}(I_1) \otimes \cdots \otimes \mathrm{Obs}^{\mathrm{cl}}(I_k) \xrightarrow{m_J^{I_1} \otimes \cdots \otimes m_J^{I_k}} \mathrm{Obs}^{\mathrm{cl}}(I) \otimes \cdots \otimes \mathrm{Obs}^{\mathrm{cl}}(I) \to \mathrm{Obs}^{\mathrm{cl}}(I),$$

where the second arrow is multiplication in the algebra $\mathrm{Obs}^{\mathrm{cl}}(I)$.

Remark 1.3. Notice that we do *not* expect general observables to have the structure of a commutative algebra. However, observables in *classical* field theory form a commutative algebra, or, in a derived setting a commutative dg algebra.

Observables in Quantum Mechanics

In a quantum setting we are forced to give up the assumption that we can simultaneously perform multiple experiments. Measurements disturb the studied system and thus, we can not expect to retain the same commutative multiplication on the observables as in the classical case.

We denote with $\mathrm{Obs}^{\mathrm{qm}}(I)$ the collection of observables corresponding to all the possible measurements one could perform on the quantum system within an open time interval $I \subset \mathbb{R}$. There is a natural map

$$\mathrm{Obs}^{\mathrm{qm}}(I) \to \mathrm{Obs}^{\mathrm{qm}}(J),$$

if $I \hookrightarrow J$ is the inclusion of a shorter time interval into a larger one. Thus, $\mathrm{Obs}^{\mathrm{qm}}$ is a precosheaf. Although we can not perform simultaneous experiments, we can combine measurements occurring at different times. More precisely, if $I_1, \ldots, I_k \subset I$ is a finite collection of disjoint open subsets of $I \subset \mathbb{R}$, we expect to have a linear map

$$m_J^{I_1,\ldots,I_k} : \mathrm{Obs}^{\mathrm{qm}}(I_1) \otimes \cdots \otimes \mathrm{Obs}^{\mathrm{qm}}(I_k) \to \mathrm{Obs}^{\mathrm{qm}}(I).$$

1.2.2 Prefactorization Algebras

We give a preliminary definition for the notion of a *prefactorization algebra*, motivated by the previously discussed properties of both classical- and quantum observables. The formal definitions are spelled out later in chapter 5.

Definition 1.1. A *prefactorization algebra* $\mathcal{F}$ of vector spaces on a topological space M is a rule that assigns a vector space $\mathcal{F}(U)$ to each open set $U \subset M$, together with the following data

- for each inclusion of open subsets $U \subset V$, we have a linear map
$$m_V^U : \mathcal{F}(U) \to \mathcal{F}(V);$$
- for every finite collection of disjoint open subsets $U_1, \ldots, U_k \subset V$ we have a linear map
$$m_V^{U_1,\ldots U_k} : \mathcal{F}(U_1) \otimes \cdots \otimes \mathcal{F}(U_k) \to \mathcal{F}(V).$$

The maps are compatible in a certain natural way. The simplest case of this compatibility is, if $U \subset V \subset W$ is a sequence of open sets, the map $\mathcal{F}(U) \to \mathcal{F}(W)$ agrees with the composition through $\mathcal{F}(V)$.

Remark 1.4. A *factorization algebra* is a prefactorization algebra that satisfies a certain *local-to-global* property, that is, we can 'glue' observables on smaller open subset to obtain observables on larger open subsets.

In this thesis we want to study the structure on the classical observables that emerge from a derived, perturbative description of the space of solutions.

To that end, assume we have given a sheaf $\mathcal{EL}$ of derived formal moduli problems on a spacetime manifold M. Let $U \subset V$ be an open subset. Then, restriction of solutions from V to U induces a natural map

$$\mathcal{EL}(V) \to \mathcal{EL}(U).$$

Since we can pullback functions along maps of spaces, we get a structure map

$$\mathrm{Obs}^{\mathrm{cl}}(U) \to \mathrm{Obs}^{\mathrm{cl}}(V).$$

Hence, $\mathrm{Obs}^{\mathrm{cl}}$ has the structure of a prefactorization algebra. Since $\mathcal{EL}$ is a sheaf, it is also a factorization algebra [CG16].

1.2.3 Observables in Gauge Theories

Working in a derived setting, we deal with factorization algebras taking values in cochain complexes, that is dg vector spaces, rather than vector spaces. Namely, the factorization algebra of observables assigns to any open subset $U \subset M$ of a spacetime M a cochain complex $(\mathrm{Obs}^\bullet(U), d)$. We give a physical interpretation of these cohomological aspects for observables in gauge theories.

The physical observables, compatible with the gauge symmetries of the theory, turn out to be elements in $H^0(\mathrm{Obs}^\bullet(U))$. Thus, physically meaningful observables can be identified with *closed* elements $O \in \mathrm{Obs}^0(U)$, that is elements obeying the equation

$$\mathrm{d}O = 0.$$

Two elements $O, O' \in \mathrm{Obs}^0(U)$ belong to the same gauge equivalence class if they differ by an exact element, that is

$$O \sim O'$$

if $O' = O + dO''$ for some $O'' \in \mathrm{Obs}^{-1}(U)$. We can interpret this equivalence relation by saying that the observables O and O' are physically indistinguishable, that is, they can not be distinguished by any performable measurement.

Remark 1.5. Giving a physical interpretation for the cohomology groups $H^n(\mathrm{Obs}^\bullet(U))$ with $n < 0$ is a more involved task that we will not elaborate here. Roughly speaking, one can think of $H^{-1}(\mathrm{Obs}^\bullet(U))$ as the group of symmetries of the trivial observable $0 \in H^0(\mathrm{Obs}^\bullet(U))$. Then, elements in $H^{-2}(\mathrm{Obs}^\bullet(U))$ can be interpreted as describing higher symmetries, that is symmetries of the identity symmetry of $0 \in H^0(\mathrm{Obs}^\bullet(U))$, and so on. We refer to [CG16] for more details.

Remark 1.6. The cohomological aspects of gauge theory are known in the physics literature as BRST formalism [FHM05].

In this thesis we study observables in 3-dimensional Chern-Simons gauge theories. The following section should provide a brief introduction in this type of Lagrangian field theory.

1.3 Chern-Simons Theory

Chern-Simons theories play a central role throughout many areas in mathematics and theoretical physics. Starting with E. Witten [Wit89b], who demonstrated that the quantization of non-abelian Chern-Simons theories leads to new topological invariants. He also explored the role of the Chern-Simons action in quantum theories of gravity in $2+1$ dimensions [Wit89a]. But Chern-Simons theories are not only interesting for their theoretical novelty, but they also play a crucial role in applications to condensed matter physics, in particular in the quantum Hall effect [Wen95], [Zee95].

Chern-Simons theory differs from most field theories as it is an example of a topological field theory. Namely, the theory is defined on an orientable manifold without the datum of a metric. Moreover, it is a gauge theory and as such arises as principal bundles equipped with the geometric data of a connection. We provide an elementary introduction into the geometric formulation of gauge theories in chapter 2. In the following, we just give a brief overview on the rich topic of Chern-Simons field theory. Good references for an extensive review are [Fre95] and [Fre02], on which also this introductory discussion is based.

1.3.1 The Chern-Simons Action

We start by introducing the Chern-Simons form defining the action functional in Chern-Simons theory. The Chern-Simons form of a connection was first introduced by S. S. Chern und J. Simons in [CS74]. The form was originally used to study secondary characteristic classes before it was reinterpreted as the Lagrangian of a field theory on compact 3-manifolds.

The Chern-Simons Form

Throughout, let M be a closed oriented 3-manifold and let $P \to M$ be a fixed principal G-bundle. A connection is characterized by a *connection 1-form* ω taking values in the Lie algebra $\mathfrak{g}$ of the structure group G

$$\omega \in \Omega^1(P; \mathfrak{g}).$$

The *curvature* Ω of the connection is the $\mathfrak{g}$-valued 2-form

$$\Omega = d\omega + [\omega \wedge \omega] \in \Omega^2(P; \mathfrak{g}),$$

where d is the exterior derivative and $[- \wedge -]$ is the bilinear product obtained by composing the wedge product with the Lie algebra bracket. In order to define the Chern-Simons form, we have to choose a symmetric bilinear form on the Lie algebra of G

$$\langle -, - \rangle : \mathfrak{g} \otimes \mathfrak{g} \to \mathbb{R},$$

that is invariant under the adjoint action of the Lie group. We can associate to this inner product a closed scalar differential form $p(\Omega) := \langle \Omega \wedge \Omega \rangle$, called *Chern-Weil form*. The form $p(\Omega)$ is invariant and horizontal and thus uniquely defines a 4-form on M whose lift is $p(\Omega)$. We denote this form on M also with $p(\Omega)$. The de Rham cohomology class of $p(\Omega)$ is independent of the connection ω and thus a topological invariant of the bundle. However, we are more interested in the geometrical aspects, which follow from the fact that $p(\Omega)$ on P is exact. More precisely, we construct the *Chern-Simons form* $\alpha(\omega)$ as the differential 3-form on P that satisfies $d\alpha(\omega) = p(\Omega)$. Explicitly, the Chern-Simons form is defined by

$$\alpha(\omega) := \langle \omega \wedge \Omega \rangle - \frac{1}{6} \langle \omega \wedge [\omega \wedge \omega] \rangle \in \Omega^3(P).$$

This form is interpreted as the *Lagrangian* of a field theory on a 3-dimensional spacetime.

The Chern-Simons Functional

The classical Chern-Simons action is defined by integrating the Lagrangian $\alpha(\omega)$ over the spacetime manifold M. However, the form $\alpha(\omega)$ lives on the total space P of the bundle rather than on M. Hence, we need a section of the bundle in order to define the action. In general, global sections do not exist, unless the bundle is *trivializable*. For simplicity, let $P \to M$ be a trivializable G-bundle and let $s : M \to P$ be a section. Moreover, we choose the inner product $\langle -, - \rangle$ on $\mathfrak{g}$ such that $\alpha(\theta)$ has integral periods, for θ the Maurer-Cartan form on G. This normalization is crucial for the gauge invariance of the action. The *Chern-Simons functional* is defined by

$$\mathcal{S}_{M,P}(s,\omega) := \int_M s^* \alpha(\omega) \pmod 1.$$

This functional is gauge invariant. Since any two sections are related by gauge transformations, it is also independent of the section s. Thus, if we let $\mathcal{A}(P)$ be the affine space of connections on P, the Chern-Simons action

$$\mathcal{S}_{M,P} : \mathcal{A}(P) \to \mathbb{R}/\mathbb{Z}$$

is well-defined and gauge invariant.

Classical Solutions

In classical field theory, we are interested in the critical points of the action functional. Since the space of connections $\mathcal{A}(P)$ is an affine space, we consider variations of $\mathcal{S}_{M,P}$ along paths $\omega_t = \omega + t\bar{\omega}$ of connections on P, where $\bar{\omega}$ is a 1-form on M with values in the adjoint bundle. Denote with $\omega = \omega_0$ and $\dot{\omega} = \frac{d}{dt}\Big|_{t=0} \omega_t$. For any section $s : M \to P$, one can show that

$$\frac{d}{dt}\Big|_{t=0} \mathcal{S}_{M,P}(s,\omega_t) = 2 \int_M s^* \langle \Omega \wedge \dot{\omega} \rangle.$$

It follows that the Chern-Simons action is constant along any path of *flat connections*, that is, connections with curvature $\Omega = 0$.

1.3.2 Abelian Chern-Simons Theory

Our main interest lies on Chern-Simons theories with an abelian group structure. Here, we assume for simplicity that all bundles are trivializable. The more general case of abelian Chern-Simons theories for *non-trivializable* bundles is discussed later in chapter 6.

Let us consider Chern-Simons theory with a $U(1)$ group structure and let $P \to M$ be a trivializable $U(1)$-bundle. Denote with $A \in \Omega^1(M;\mathbb{R})$ and $F \in \Omega^2(M;\mathbb{R})$ the representatives of the connection and curvature on the spacetime manifold. The Chern-Simons action takes the following simple form

$$\mathcal{S}_{M,P}(A) = \frac{k}{4\pi}\int_M A \wedge dA \pmod 1, \quad k \in \mathbb{Z}.$$

The factor in front of the integral is due to the normalization of the inner product on the Lie algebra. Variation of the action leads to the Euler-Lagrange equation

$$F = dA = 0.$$

Hence, the classical moduli space of solutions is given by the space of flat $U(1)$ connections over M modulo gauge equivalences. We recover the same result when we treat the general case of non-trivializable $U(1)$-bundles in chapter 6.

Remark 1.7. The Chern-Simons action for any abelian Lie group G is of the same form as the above functional. More precisely, the Chern-Simons form of a connection $A \in \Omega^1(M;\mathfrak{g})$ on the spacetime manifold M is given by

$$\langle A \wedge dA \rangle,$$

where $\langle -, - \rangle$ is the appropriate invariant pairing on the abelian Lie algebra $\mathfrak{g}$ of G.

Example 1.1. The quantum Hall effect is an interesting example from condensed matter physics that involves action functionals of the above type. For instance, let us consider the *integer* quantum Hall effect. We can use the $U(1)$ Chern-Simons theory to explain why we expect a quantization of the Hall conductivity in the first place. To that end, take $a \in \Omega^1(M;\mathbb{R})$ to be a non-dynamical background gauge field of electromagnetism.

From the viewpoint of effective field theory, we take the Chern-Simons action to be the *effective action*, describing perturbations around a given Hall state

$$\mathcal{S}_{\text{eff}}(a) = \frac{k}{4\pi}\int_M a \wedge da \quad (\text{mod } 1), \quad k \in \mathbb{Z}.$$

As before, the integer k is a consequence of the gauge invariance of the action in $\mathbb{R}/\mathbb{Z}$. Consequently, the Hall current

$$J_i^{\text{Hall}} = \frac{\delta \mathcal{S}_{\text{eff}}(a)}{\delta a_i} = \epsilon_{ij}\frac{k}{2\pi}E_j$$

is quantized by the integers $k \in \mathbb{Z}$ [Wit16]. A review on the Chern-Simons approach to the quantum Hall effect can be found in [Wen95] and [Zee95].

1.4 Overview of this Thesis

In chapter 2 we give an outline on the geometrical concepts involved in gauge theory. We recall the basics of principal bundles, Lie groups and their associated Lie algebras. We then spend some time to review the notion of principal bundles equipped with connections, since they play the role of the dynamical fields in Chern-Simons theories.

Chapter 3 mainly serves as a review on aspects in homological algebra used in the scope of this thesis. In particular, we introduce notions regarding differential graded algebras, such as homotopies between dg algebra morphisms. Those are used to construct the homotopy action of the gauge group on the classical observables.

Chapter 4 deals with the mathematical framework to describe the perturbative aspects of field theory. Since we are working in a derived setting, we explain concepts in derived deformation theory that allow to study the local structure of the derived moduli space of solutions and clarify the meaning of a 'derived formal moduli problem'. Moreover, we state the main theorem that tells us to trade moduli problems with dg Lie algebras.

In chapter 5 we focus on factorization algebras, their explicit definition and some simple examples. Moreover, the factorization algebra associated to a derived formal moduli problem in classical field theory is introduced. This is the type of factorization algebra encountered in the study of observables in abelian Chern-Simons theories.

The final chapter 6 of this thesis is devoted to our main project, the construction of a homotopy action of the gauge group on the factorization algebra of observables in abelian Chern-Simons theories. The main focus lies on Chern-Simons theory with a $U(1)$ group structure. In this process, we describe the action functional for $U(1)$-bundles and its critical points, which turn out to be the flat connections. We give the derived formal moduli problems deforming flat abelian bundles and observe that the classical factorization algebra detects only the underlying Lie algebra, rather than the associated Lie group. Finally, going beyond a purely perturbative treatment, we construct a gauge transformation action on the factorization algebra of observables. The resulting equivariant structure on the observables incorporate parts of the topology of the underlying Lie group.

2 Principal Bundles and Gauge Theory

2.1 Physical Motivation

In many areas of theoretical physics we encounter fields defined on a spacetime M taking values in some other space F. For instance, presume F as a real finite-dimensional vector space, then $\phi : M \to F$ is a vector field. More generally, we can consider a family of spaces $\{F_x\}_{x \in M}$ varying over the points on M, that is $\phi(x) \in F_x$ for each $x \in M$. A field ϕ is then understood as a *section* from the spacetime manifold into the bundle of spaces over M. This is exactly the idea encoded in the mathematical theory of *fiber bundles*. Namely, fiber bundles provide a tool to describe the global structure of physical fields.

We turn back for a moment to the case of a fixed target space F over M. The corresponding bundle is called a *trivial* fiber bundle and a field is essentially the same as the data of a global function $\phi : M \to F$. Schematically, we have the following picture

$$\begin{array}{c} M \times F \\ {\scriptstyle \mathrm{pr}_1}\big\downarrow \big\uparrow{\scriptstyle \phi} \\ M \end{array}$$

where $\mathrm{pr}_1 \circ \phi = \mathrm{id}_M$. Locally, every fiber bundle can be given this product structure, meaning that infinitesimally no global twist of the fields is visible. All the *fibers*, that is the family of target spaces $\{F_x\}_{x \in M}$, are equivalent to a common space F. However, the crucial point is that the fibers have automorphisms. This means that the way they are equivalent to the common space F can vary over the spacetime manifold, encoding the global structure of the physical fields.

One particular type of bundle is the *principal G-bundle*, whose fiber is a Lie group G. Principal bundles are of enormous importance in many fields of

C. Keller, *Chern-Simons Theory and Equivariant Factorization Algebras*, BestMasters, https://doi.org/10.1007/978-3-658-25338-7_2

theoretical physics, especially for describing gauge theories in a geometrical setting. In this chapter we first give a brief overview on bundles in general, presenting the most important definitions and basic concepts. We then review the notion of principal bundles with an emphasis on their role in gauge theory. In doing so, we equip the principal bundles with the geometric data of a connection. We show that the local description of connections leads to the notion of *gauge fields* as used in physics. Moreover, we introduce the curvature of a connection, known in the physics literature as the *gauge field strength.*

The main reference for this chapter is the book by C. Isham [Ish13], presenting various topics in differential geometry and their application to modern theoretical physics, as well as the book by H. Baum [Bau14], discussing the mathematical concepts of gauge theory. In the following, proofs are often omitted and we refer to [Ish13], [Bau14] and [KN96] for more details.

2.2 Fiber Bundles

Definition 2.1. Let E, M, and F be topological spaces and $\pi : E \to M$ a continuous surjection. A *fiber bundle* is a tuple (E, M, π) subjected to the condition of *local triviality*, i.e. we require that for every $x \in M$ there exists an open neighborhood $U \subset M$ and a homeomorphism $\psi : \pi^{-1}(U) \to U \times F$, such that the following diagram is commutative

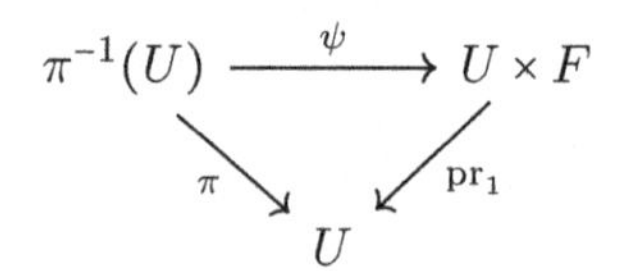

Remark 2.1. The space E is referred to as the *total space* and the space M as the *base space* of the bundle. The continuous surjection $\pi : E \to M$ is called the *projection* and F is called the *fiber* of the bundle.

Remark 2.2. A *smooth* fiber bundle is a fiber bundle where the spaces E, M, F are smooth manifolds, the projection π is a smooth surjection and the inverse images $\pi^{-1}(\{x\})$ are all diffeomorphic to the fiber F.

Definition 2.2. Let (E, M, π) and $(\tilde{E}, \tilde{M}, \tilde{\pi})$ be fiber bundles. A *bundle map* is a pair (ϑ, χ), where $\vartheta : E \to \tilde{E}$ and $\chi : M \to \tilde{M}$, such that the following diagram is commutative

$$\begin{array}{ccc} E & \xrightarrow{\vartheta} & \tilde{E} \\ {\scriptstyle\pi}\downarrow & & \downarrow{\scriptstyle\tilde{\pi}} \\ M & \xrightarrow{\chi} & \tilde{M} \end{array}$$

Remark 2.3. Definition 2.2 implies that for all $x \in M$ the map ϑ restricts to a map $\pi^{-1}(\{x\}) \to \tilde{\pi}^{-1}(\{\chi(x)\})$, in other words ϑ is fiber-preserving.

Definition 2.3. A fiber bundle (E, M, π) with fiber F is called *trivializable* if it is isomorphic to the product bundle $(M \times F, M, \mathrm{pr}_1)$, where pr_1 is the projection onto the first factor.

Roughly speaking, *global sections* assign to each point in the base space an element of the fiber. However, global sections do not always exist, but demanding the bundles to be locally trivial ensures the existence of *local sections*. The following definition formalizes these notions.

Definition 2.4. A *local section* of a fiber bundle (E, M, π) is a map

$$s_U : U \to E,$$

for an open subset $U \subset M$, such that the image of each point $x \in U$ lies in the fiber $\pi^{-1}(\{x\})$ over x. More precisely, we have

$$\pi \circ s_U = \mathrm{id}_U.$$

Similarly, a *global section* is a map $s : M \to E$ such that $\pi \circ s = \mathrm{id}_M$.

2.2.1 Vector Bundles

If the fiber of a bundle is equipped with the structure of a real vector space, the resulting bundle is called a *vector bundle*. The space of sections of a vector bundle carries a natural structure of a vector space, generalizing the idea of the linear space of functions on a manifold.

Definition 2.5. A *real vector bundle* of dimension k is a fiber bundle (E, M, π) in which each fiber has the structure of a k-dimensional real vector space. Moreover, we require that for every $x \in M$ there exists an open neighborhood $U \subset M$ and a local trivialization $\psi : \pi^{-1}(U) \to U \times \mathbb{R}^k$ such that the restriction $\psi|_{\pi^{-1}(\{y\})} : \pi^{-1}(\{y\}) \to \{y\} \times \mathbb{R}^k$ is a linear isomorphism for all $y \in U$.

Definition 2.6. Let (E, M, π) and $(\tilde{E}, \tilde{M}, \tilde{\pi})$ be two vector bundles. A *vector bundle map* is a bundle map (ϑ, χ), in which the restriction of $\vartheta : E \to \tilde{E}$ to each fiber is a linear map.

Remark 2.4. The space of sections of a vector bundle (E, M, π) carries a natural structure of a real vector space. Indeed, let $s, \tilde{s} : M \to E$ be two sections. Then, we can define

$$\begin{aligned}(s + \tilde{s})(x) &\coloneqq s(x) + \tilde{s}(x) \\ (rs)(x) &\coloneqq rs(x)\end{aligned}$$

for all $x \in M$ and $r \in \mathbb{R}$.

2.3 Principal Fiber Bundles

A *principal fiber bundle* is a bundle whose fibers are diffeomorphic to a Lie group G. More precisely, the total space carries an action of the Lie group G, turning each fiber into a right G-torsor. The essential definitions are the following.

Definition 2.7. Let P and M be smooth manifolds, $\pi : P \to M$ a smooth surjection and G a Lie group. A *principal G-bundle* is a fiber bundle (P, M, π, G), together with a right G-action

$$\begin{aligned}\Gamma : P \times G &\to P \\ (p, g) &\mapsto R_g(p) = p.g\end{aligned}$$

such that

- G preserves the fibers and acts freely and transitively on them;
- there exist a *G-equivariant* local trivialization of the bundle.

Remark 2.5. The manifold P is referred to as the *total space* and the manifold M as the *base space* of the bundle. The smooth surjection $\pi : P \to M$ is called the *projection* and the Lie group G is called the *structure group*.

Remark 2.6. G-equivariant local triviality requires that for each $x \in M$ there exists an open neighborhood $U \subset M$ and a diffeomorphism $\psi : \pi^{-1}(U) \to U \times G$, such that the following diagram is commutative

$$\begin{array}{ccc} \pi^{-1}(U) & \xrightarrow{\psi} & U \times G \\ & {\scriptstyle \pi}\searrow \quad \swarrow{\scriptstyle \mathrm{pr}_1} & \\ & U & \end{array}$$

Hence, we can write $\psi(p) = (\pi(p), \phi(p))$ for some G-equivariant diffeomorphism

$$\phi : \pi^{-1}(U) \to G, \quad \text{satisfying} \quad \phi(p.g) = \phi(p)g.$$

On non-empty overlaps $U_\alpha \cap U_\beta$, we have two trivializing maps $\psi_\alpha(p) = (\pi(p), \phi_\alpha(p))$ and $\psi_\beta(p) = (\pi(p), \phi_\beta(p))$ for each $p \in \pi^{-1}(\{x\})$, where $x \in U_\alpha \cap U_\beta$. From the G-equivariance of ϕ it follows that there has to be a map

$$\phi_\alpha(p)\phi_\beta(p)^{-1} : G \to G,$$

which is independent of the choice of $p \in \pi^{-1}(\{x\})$. We can identify this map with an element of G, more precisely, we can define *transition functions* $\phi_{\alpha\beta}$ on each non-empty overlap $U_\alpha \cap U_\beta$

$$\phi_{\alpha\beta} : U_\alpha \cap U_\beta \to G, \quad \phi_{\alpha\beta}(x) := \phi_\alpha(p)\phi_\beta(p)^{-1},$$

for any $p \in \pi^{-1}(\{x\})$. Notice that in physics literature the transition functions are often referred to as *local gauge transformations*.

Definition 2.8. Let (P, M, π, G) and $(\tilde{P}, \tilde{M}, \tilde{\pi}, G)$ be two principal G-bundles. A *principal bundle map* is a bundle map (ϑ, χ), where ϑ is G-equivariant in the sense that

$$\vartheta(p.g) = \vartheta(p).g,$$

for all $p \in P$ and $g \in G$.

Remark 2.7. We denote with $\mathbf{Bun}^G(M)$ the category whose objects are principal G-bundles over M and whose morphisms are principal bundle maps.

Lemma 2.1. *Let (ϑ, id_M) be a principal bundle map between the principal bundles (P, M, π) and $(\tilde{P}, M, \tilde{\pi})$, covering the identity on M. Then ϑ is an isomorphism.*

Remark 2.8. It follows from lemma 2.1 that the category $\mathbf{Bun}^G(M)$ is a groupoid.

Recall that a bundle (P, M, π, G) is trivializable if it is isomorphic to the product bundle $(M \times G, M, \mathrm{pr}_1, G)$. Lemma 2.2 states that trivializability of a principal bundle is equivalent to the existence of a global section in the bundle. Hence, in general, principal bundles do not admit global sections unless they are trivial.

Lemma 2.2. *A principal G-bundle is* trivializable *if and only if it admits a global section.*

Remark 2.9. Since we require local triviality in our definition for principal bundles it is guaranteed that local sections exist. In fact, there are local sections $s_U : U \to \pi^{-1}(U)$ canonically associated to the local trivialization $\psi : \pi^{-1}(U) \to U \times G$, defined so that for every $x \in U$ we have $\psi(s_U(x)) = (x, e)$.

Remark 2.10. Simply connected Lie groups play a special role in theories involving the use of principal bundles. Namely, for G a simply connected Lie group and M a manifold of dimension $d \leq 3$, any principal bundle (P, M, π, G) admits a global section and hence is trivializable.

2.3.1 Basics of Lie Groups and Lie Algebras

In discussing principal bundles we have to deal with some basic technology and results from the theory of Lie groups and their associated Lie algebras. Therefore, this section gives a short outline on this topic and introduces the main tools and notions that are needed later on.

Lie Groups

Recall that Lie groups are groups which are also differentiable manifolds so that the group operations are smooth. Lie groups are of enormous significance in describing continuous symmetries of mathematical objects and are therefore encountered in many areas of modern theoretical physics and mathematics.

Definition 2.9. A *real Lie group* is a group G that is a differentiable manifold in such a way that the following maps are smooth

– (*Group multiplication*)

$$\mu : G \times G \to G \ , \ \mu((g, \tilde{g})) = g\tilde{g};$$

– (*Inverse element*)

$$i : G \to G \ , \ i(g) = g^{-1}.$$

Definition 2.10. Let G be a Lie group. For every $g \in G$ we define *left translation* as the map

$$l_g : G \to G, \quad h \mapsto gh,$$

and similarly, we define *right translation* as the map

$$r_g : G \to G \quad h \mapsto hg,$$

for all $h \in G$.

Remark 2.11. Every $g \in G$ defines a smooth map $\mathrm{Ad}_g : G \to G$ by $\mathrm{Ad}_g = l_g \circ r_{g^{-1}}$, called the *adjoint map.* That is

$$\mathrm{Ad}_g(h) = ghg^{-1},$$

for all $h \in G$.

Example 2.1. The Lie group $U(n)$, called the *unitary group*, is defined by

$$U(n) := \{A \in \mathrm{GL}(n, \mathbb{C}) \mid AA^\dagger = \mathbb{1}\},$$

where $\mathrm{GL}(n, \mathbb{C})$ is the complex general linear group in n-dimensions, i.e. the Lie group of complex $n \times n$ matrices with non-zero determinant. $U(n)$ is a compact Lie group with real dimension n^2. For $n = 1$ we get the circle $U(1) := \{z \in \mathbb{C} \mid |z| = 1\}$. This is a 1-dimensional real Lie group with group multiplication given by the multiplication law for complex numbers. It is a compact, connected Lie group but it is not simply connected.

Lie Algebras

One can approach the study of Lie groups by means of their associated Lie algebras which algebraically encode parts of the Lie group's geometry. They describe how Lie groups 'look like' locally, however in general failing to capture global topological features. We review the basic notions of Lie algebras associated to Lie groups. Moreover, we show that one can use *left invariant* vector fields on the Lie group G to identify the tangent space of G at the identity element with its Lie algebra.

Definition 2.11. A vector field X on a Lie group G is *left invariant* if

$$l_{g*}X_h = X_{l_g(h)} = X_{gh},$$

for all $g, h \in G$.

Remark 2.12. The set of all left invariant vector fields on a Lie group G is denoted by $L(G)$. Very often we also adapt the notation $L(G) = \mathfrak{g}$.

Remark 2.13. The set of left invariant vector fields on G is a real vector space. Furthermore, given two left invariant vector fields X_1 and X_2 on G, their commutator is again a left invariant vector field

$$\begin{aligned} l_{g*}[X_1, X_2] &= [l_{g*}X_1, l_{g*}X_2] \\ &= [X_1, X_2]. \end{aligned}$$

Thus, $X_1, X_2 \in L(G)$ implies $[X_1, X_2] \in L(G)$ and we call the set $L(G)$ the *Lie algebra* of G.

Lemma 2.3. *There is an isomorphism of the vector space $L(G)$ of left invariant vector fields on the Lie group G with the tangent space T_eG at the identity element $e \in G$*

$$T_eG \xrightarrow{\cong} L(G), \quad A \mapsto L^A,$$

where we define the left-invariant vector field on G by

$$L_g^A := l_{g*}A,$$

for all $g \in G$ and $A \in T_eG$.

Remark 2.14. $L(G)$ is a vector space with dimension $\dim(T_eG) = \dim(G)$.

Remark 2.15. We can identify the Lie algebra of G with T_eG. The Lie bracket on T_eG is constructed from the commutator on the left invariant vector fields $L(G)$ using the isomorphism described in lemma 2.3. Explicitly, the Lie bracket of two elements $A, B \in T_eG$ is defined as

$$[A, B] := [L^A, L^B]_e.$$

A key result in the theory of Lie groups is that every left invariant vector field on a Lie group is complete. We use this result to introduce the *exponential map* from the Lie algebra $L(G)$ to the Lie group G. It allows to reconstruct the group structure locally.

Definition 2.12. The *exponential map* $\exp : T_eG \to G$ is defined for $A \in T_eG$ by

$$\exp A := \exp tA\Big|_{t=1},$$

where $t \to \exp tA$, for all $t \in \mathbb{R}$, is the unique integral curve of the left invariant vector field L^A, passing at $t = 0$ through $e \in G$.

Remark 2.16. Recall from remark 2.11 that for each $g \in G$ we have the adjoint map $\mathrm{Ad} : G \to G$, preserving the identity. Thus, we get a linear representation of the group G on its Lie algebra T_eG, known as the *adjoint representation* $\mathrm{ad}_g = \mathrm{Ad}_{g_*} : T_eG \to T_eG$, for all $g \in G$. The representation is defined on all elements $X \in T_eG$ by

$$\mathrm{ad}_g(X) = \frac{d}{dt} g \exp(tX) g^{-1}\Big|_{t=0}.$$

Transformation Groups

We now turn to Lie groups that are of special interest for our discussion, namely groups that act on a space via automorphisms. We first recall the definition of a right group action on a manifold and then introduce the notion a *fundamental vector fields* on a manifold induced by the Lie algebra $L(G)$.

Definition 2.13. A *right action* of a Lie group G on a differentiable manifold P is a homomorphism $g \mapsto R_g$ from G into the group of diffeomorphisms $\mathrm{Diff}(P)$ with the property that the map $\Gamma : P \times G \to P$, defined by

$$\Gamma : P \times G \to P, \quad (p, g) \mapsto R_g(p) = p.g,$$

is smooth.

Definition 2.14. Let G be a Lie group together with a right action $\Gamma : P \times G \to P$ on a manifold P. This action induces a map $L(G) \to \mathfrak{X}(P)$, assigning to every $A \in L(G)$ the *fundamental vector field* $\tilde{X}^A$ on P defined by

$$\tilde{X}^A_p := \frac{d}{dt} p.\exp tA\Big|_{t=0}.$$

Remark 2.17. Alternatively, we can define the fundamental vector field induced by $A \in T_eG$ as

$$\tilde{X}_p^A := \mathrm{P}_{p*}A,$$

where $\mathrm{P}_p(g) := \Gamma(p,g) = p.g$.

The following theorem guarantees that the map of definition 2.14 is a *morphism of Lie algebras*. In other words, the vector fields on the manifold P represent the Lie algebra of G homomorphically.

Theorem 2.1. *Let P be a manifold on which a Lie group G has a right action. Then the map $A \mapsto \tilde{X}^A$, which associates to each $A \in T_eG$ the fundamental vector field $\tilde{X}^A \in \mathfrak{X}(P)$, is a homomorphism of $L(G) \simeq T_eG$ into the Lie algebra of all vector field on P, i.e.*

$$[\tilde{X}^A, \tilde{X}^B] = \tilde{X}^{[A,B]},$$

for all $A, B \in T_eG$.

The Maurer-Cartan Form

Every Lie group carries a canonical 1-form, the *Maurer-Cartan* form, defined globally on the Lie group. The Maurer-Cartan form defines a linear map of the tangent space at each element of the Lie group into its Lie algebra and thus entails infinitesimal information about the group structure.

Definition 2.15. The *Maurer-Cartan* form is the $L(G)$-valued 1-form θ on G defined by

$$\theta_g = l_{g^{-1}*} : T_gG \to T_eG,$$

for all $g \in G$.

Remark 2.18. In other words, θ associates with any $v \in T_gG$ the left invariant vector field on G whose value at $g \in G$ is precisely the given tangent vector v.

Remark 2.19. The Maurer-Cartan form θ is *left invariant*, i.e.

$$l_g^*(\theta_h) = \theta_{g^{-1}h}$$

for all $g, h \in G$. Moreover, it satisfies the *Maurer-Cartan structure equation*

$$d\theta = -\frac{1}{2}[\theta \wedge \theta],$$

where $[- \wedge -]$ is the bilinear product obtained by composing the wedge product of the two $L(G)$-valued 1-forms with the bracket of the Lie algebra $L(G)$.

More generally, for any smooth manifold M we can define a bilinear product $[- \wedge -]$ on the space $\Omega^\bullet(M, \mathfrak{g})$ of Lie algebra-valued differential forms by

$$[\alpha \otimes x \wedge \beta \otimes y] \coloneqq \alpha \wedge \beta \otimes [x, y],$$

for $x, y \in \mathfrak{g}$ and $\alpha, \beta \in \Omega^\bullet(M, \mathbb{R})$.

2.4 Connections on Principal Bundles

In general, a connection is a mathematical tool that enables to define the concept of parallel translation on a bundle. Thus, a connection should provide a natural way to compare and connect points in adjacent fibers. Here, we give three equivalent characterizations, namely connections as horizontal distributions, as Lie algebra-valued 1-forms on the total space and as their local representatives.

2.4.1 Connections as Horizontal Distributions

Motivated by the idea of parallel translation, we note that a connection should provide a consistent way of moving from fiber to fiber through the bundle. This suggests that we should look for vector fields whose flow lines point from one fiber to another. This idea will be made precise by introducing connections as horizontal distributions. In the following, let (P, M, π, G) be a principal G-bundle and let $x \in M$ and $p \in \pi^{-1}(\{x\})$.

Definition 2.16. Let T_pP denote the tangent space at the point $p \in P$. The *vertical subspace* V_pP of T_pP is defined by

$$V_pP \coloneqq \{v \in T_pP \mid \pi_* v = 0\}.$$

Remark 2.20. Recall that for each $A \in \mathfrak{g}$ we can assign the fundamental vector field $\tilde{X}^A$ on P that represents the Lie algebra $\mathfrak{g}$ homomorphically. A vector $\tilde{X}_p^A$ is tangent to the fiber and thus belongs to the vertical subspace V_pP. Indeed, by definition 2.14 we have

$$\begin{aligned} \pi_* \tilde{X}_p^A &= \frac{d}{dt}\pi(p.\exp tA)\Big|_{t=0} \\ &= \frac{d}{dt}\pi(p)\Big|_{t=0} \\ &= 0. \end{aligned}$$

We are interested in constructing vectors that point away from the fibers rather than along them. In other words, we are looking for vectors that live in a space complementary to the vertical subspace V_pP. However, per se there is no natural choice for such complement spaces. This is exactly what a connection provides, a consistent way of assigning a *horizontal subspace* at each point of the total space.

Definition 2.17. A *connection* on a principal bundle (P, M, π, G) is a smooth assignment of a subspace H_pP of T_pP to each point $p \in P$ such that

- $T_pP \simeq V_pP \oplus H_pP$;
- $R_{g_*} H_pP = H_{p.g}P$,

for all $p \in P$ and $g \in G$. In other words, a connection is a G-invariant distribution $H \subset TP$ complementary to the vertical distribution V.

2.4.2 Connections as Lie Algebra-valued 1-Forms

There is an alternative to characterize connections, namely by means of certain 1-forms on P with values in the Lie algebra $\mathfrak{g}$.

Definition 2.18. A *connection 1-form* on a principal bundle (P, M, π, G) is a 1-form $\omega \in \Omega^1(P; \mathfrak{g})$ satisfying the following properties:

- $\omega_p(\tilde{X}^A) = A$, for all $p \in P$ and $A \in \mathfrak{g}$;
- $R_g^* \omega = \mathrm{ad}_{g^{-1}} \circ \omega$, for all $g \in G$.

Remark 2.21. We denote with $\mathbf{Bun}_\omega^G(M)$ the category whose objects are principal G-bundles over M equipped with a connection and whose morphisms are principal bundle maps.

Remark 2.22. Let $\omega \in \Omega^1(P; \mathfrak{g})$ be a 1-form as given in definition 2.18. Then the distribution $H = \ker(\omega)$ defines a connection on P.

2.4.3 Local Representatives of the Connection 1-Form

Recall that we have canonical local sections $s_\alpha : U_\alpha \to \pi^{-1}(U_\alpha)$ associated to the local trivialization of the principal bundle. Thus, we can pullback the connection 1-form ω to obtain a *local representative* of the connection 1-form on the base manifold M

$$A_\alpha := s_\alpha^* \omega \in \Omega^1(U_\alpha; \mathfrak{g}).$$

Local representatives are often referred to as *gauge fields* in the physics literature. Since ω is defined globally, we have $\omega_\alpha = \omega_\beta$ on $\pi^{-1}(U_\alpha \cap U_\beta)$. The following lemma shows how the corresponding local representatives are related on $U_\alpha \cap U_\beta$.

Lemma 2.4. *Let $\omega \in \Omega^1(P; \mathfrak{g})$ be a connection on a principal bundle (P, M, π, G) and let $s_\alpha : U_\alpha \to \pi^{-1}(U_\alpha)$ and $s_\beta : U_\beta \to \pi^{-1}(U_\beta)$ be two local sections on $U_\alpha, U_\beta \subset M$. Moreover, let $\phi_{\alpha\beta} : U_\alpha \cap U_\beta \to G$ be the unique transition function such that $s_\beta(x) = s_\alpha(x).\phi_{\alpha\beta}(x)$, for all $x \in U_\alpha \cap U_\beta$. Denote with A_α and A_β the local representatives of ω with respect to s_α and s_β. Then for $U_\alpha \cap U_\beta \neq 0$ the two local representatives are related by*

$$A_\beta = ad_{\phi_{\alpha\beta}^{-1}} \circ A_\alpha + \phi_{\alpha\beta}^* \theta,$$

where θ is the Maurer-Cartan form on G.

Proof. Let $x \in U_\alpha \cap U_\beta$ and $X \in T_x M$. First, use the relation $s_\beta(x) = s_\alpha(x).\phi_{\alpha\beta}(x)$ to factorize the local section s_β as

$$\begin{aligned} U_\alpha \cap U_\beta \xrightarrow{s_\alpha \times \phi_{\alpha\beta}} \pi^{-1}(U_{\alpha\beta}) \times G &\xrightarrow{\Gamma} \pi^{-1}(U_{\alpha\beta}) \\ x \mapsto (s_\alpha(x), \phi_{\alpha\beta}(x)) &\mapsto s_\alpha(x).\phi_{\alpha\beta}, \end{aligned}$$

where Γ denotes the right action of G on P. We get

$$\begin{aligned} (A_\beta)_x(X) &= ((s_\alpha \times \phi_{\alpha\beta})^* \Gamma^* \omega)_x(X) \\ &= (\Gamma^* \omega)_{(s_\alpha(x), \phi_{\alpha\beta}(x))}(s_{\alpha *} X, \phi_{\alpha\beta *} X). \end{aligned}$$

We now make use of the following isomorphism, given here for a general product manifold $U \times V$

$$T_rU \oplus T_sV \simeq T_{(r,s)}(U \times V)$$
$$(\bar{u}, \bar{v}) \mapsto i_{s*}\bar{u} + j_{r*}\bar{v},$$

with the injections $i_s : U \to U \times V$, $i_s(x) := (x, s)$, and $j_r : V \to U \times V$, $j_r(y) := (r, y)$, respectively, defined for each $r \in U$ and $s \in V$. Define functions $i_g : P \to P \times G$, $i_g(p) := (p, g)$, and $j_p : G \to P \times G$, $j_p(g) := (p, g)$, for all $p \in P$ and $g \in G$. We can re-express the local representative as

$$\begin{aligned}(A_\beta)_x(X) &= \omega_{s_\alpha(x).\phi_{\alpha\beta}(x)}((\Gamma \circ i_{\phi_{\alpha\beta}(x)})_* s_{\alpha*}X + (\Gamma \circ j_{s_\alpha(x)})_* \phi_{\alpha\beta_*}X) \\ &= (R^*_{\phi_{\alpha\beta}(x)}\omega)_{s_\alpha(x)}(s_{\alpha*}X) + \omega_{s_\alpha(x).\phi_{\alpha\beta}(x)}((\Gamma \circ j_{s_\alpha(x)})_* \phi_{\alpha\beta_*}X)\end{aligned}$$

where $\Gamma \circ i_g = R_g$ denotes the right action of the element $g \in G$ on P. Now, since ω is a connection 1-form we have

$$\begin{aligned}(R^*_{\phi_{\alpha\beta}(x)}\omega)_{s_\alpha(x)}(s_{\alpha*}X) &= \mathrm{ad}_{\phi_{\alpha\beta}(x)^{-1}} \circ \omega_{s_\alpha(x)}(s_{\alpha*}X) \\ &= \mathrm{ad}_{\phi_{\alpha\beta}(x)^{-1}} \circ (s^*_\alpha\omega)_x(X) \\ &= \mathrm{ad}_{\phi_{\alpha\beta}(x)^{-1}} \circ (A_\alpha)_x(X).\end{aligned}$$

We know that $\phi_{\alpha\beta_*}X = L^Y_{\phi_{\alpha\beta}(x)} = l_{\phi_{\alpha\beta}(x)_*}Y$ is a left invariant vector field for some $Y \in T_eG$. More precisely, according to the definition of the Maurer-Cartan form we have $Y = \theta_{\phi_{\alpha\beta}(x)}(\phi_{\alpha\beta_*}X)$. Furthermore, notice that $\Gamma \circ j_p := \mathrm{P}_p : G \to P$, $g \mapsto p.g$, defines the right action of the group G for every $p \in P$ and we can use that the left invariant vector field L^Y on G and the induced fundamental vector field $\tilde{X}^Y$ on P are P_p-related for each $p \in P$, that is

$$\mathrm{P}_{p*}L^Y_g = \tilde{X}^Y_{p.g},$$

for all $g \in G$. Let us apply these observations to the case at hand

$$\begin{aligned}\omega_{s_\alpha(x).\phi_{\alpha\beta}(x)}(P_{s_\alpha(x)_*}\phi_{\alpha\beta_*}X) &= \omega_{s_\alpha(x).\phi_{\alpha\beta}(x)}(P_{s_\alpha(x)_*}L^Y_{\phi_{\alpha\beta}(x)}) \\ &= \omega_{s_\alpha(x).\phi_{\alpha\beta}(x)}(\tilde{X}^Y_{s_\alpha(x).\phi_{\alpha\beta}(x)}) \\ &= \theta_{\phi_{\alpha\beta}(x)}(\phi_{\alpha\beta_*}X) \\ &= (\phi^*_{\alpha\beta}\theta)_x X\end{aligned}$$

where we used that ω is a connection 1-form, i.e. we have $\omega(\tilde{X}^Y) = Y$. Putting everything together we find

$$(A_\beta)_x(X) = \mathrm{ad}_{\phi_{\alpha\beta}(x)^{-1}} \circ (A_\alpha)_x(X) + (\phi^*_{\alpha\beta}\theta)_x X.$$

□

Remark 2.23. For G a matrix Lie group, the relation between local representatives stated in lemma 2.4 assumes the following simpler form

$$A_\beta = \phi^{-1}_{\alpha\beta} A_\alpha \phi_{\alpha\beta} + \phi_{\alpha\beta} d\phi_{\alpha\beta}.$$

Remark 2.24. Given an open cover $\{U_\alpha\}_{\alpha\in A}$ and a family of 1-forms $A_\alpha \in \Omega^1(U_\alpha; \mathfrak{g})$ satisfying the relation in lemma 2.4, one can construct a globally defined connection 1-form $\omega \in \Omega^1(P; \mathfrak{g})$. Thus, the local representatives provide an alternative way to characterize connections on a principal bundle.

2.4.4 The Affine Space of Connections

Theorem 2.2. *Every principal G-bundle admits a connection.*

Since the theorem guarantees the existence of connections, we can raise the question of how to describe the space of connections on a principal bundle. Before giving the answer, we have to settle some terminology. First, we define the notion of a fiber bundle *associated* to a given principal bundle. Then we introduce the so-called *basic* forms on a principal bundle. Both notions are used in lemma 2.6 to state the key identification between forms on the total- and the base space that allows to characterize the space of connections.

Associated Fiber Bundles

The associated fiber bundle construction is a method to build a wide variety of fiber bundles that are related with a given principal bundle in a specific way.

Definition 2.19. Let (P, M, π, G) be a principal G-bundle, F a space on which G acts from the left via automorphisms and let $\rho : G \to \mathrm{Aut}(F)$ be the corresponding representation. Define the quotient

$$P \times_{(G,\rho)} F := (P \times F)/\sim,$$

where the equivalence relation $\sim$ is generated by the G-action

$$(p, f) \sim (p, f).g = (p.g, \rho(g^{-1}).f).$$

There is a canonical projection

$$\tilde{\pi} : P \times_{(G,\rho)} F \to M;\ \tilde{\pi}([p, f]) := \pi(p),$$

turning $(P \times_{(G,\rho)} F, M, \tilde{\pi}, F)$ into a fiber bundle with fiber F, which is said to be *associated* with the principal G-bundle (P, M, π, G) via the representation ρ.

Remark 2.25. Let F be a space with a trivial G action, i.e. $\rho(g^{-1}).f = f$ for all $g \in G$ and $f \in F$. Then, the associated bundle is canonically isomorphic to the trivial bundle $(M \times F, M, \mathrm{pr}_1)$. The isomorphism is given by $[p, f] \mapsto (\pi(p), f)$.

Example 2.2. Let F be a vector space and ρ a linear representation, $\mathrm{Aut}(F) \simeq \mathrm{GL}(F)$, then the associated bundle is a vector bundle. In particular, taking F to be the Lie algebra $\mathfrak{g}$ together with the adjoint representation $\mathrm{ad} : G \to \mathrm{GL}(\mathfrak{g})$ we obtain the *adjoint bundle* denoted by $\mathrm{ad}(P) := P \times_{G,\mathrm{ad}} \mathfrak{g}$.

Example 2.3. Let F be a smooth manifold and let G act on F via diffeomorphisms, $\mathrm{Aut}(F) = \mathrm{Diff}(F)$. In particular, taking F to be the Lie group G together with the adjoint representation $\mathrm{Ad} : G \to \mathrm{Diff}(G)$ we obtain the *adjoint principal bundle* denoted by $\mathrm{Ad}(P) := P \times_{G,\mathrm{Ad}} G$.

An important observation is that sections of an associated bundle can be completely described in terms of functions on the corresponding principal bundle.

Lemma 2.5. *Let (P, M, π, G) be a principal G-bundle and $E := P \times_{(G,\rho)} F$ the fiber bundle associated to the principal bundle via $\rho : G \to \mathrm{Aut}(F)$. Sections $\Gamma(M, E)$ of the associated bundle are in one-to-one correspondence with maps*

$$\gamma : P \to F,$$

satisfying the following equivariance property

$$\gamma(p.g) = \rho(g^{-1}).\gamma(p).$$

Remark 2.26. We denote the space of equivariant smooth maps by $C^\infty(P,F)^{(G,\rho)}$.

Basic Forms

In the following, let (P,M,G,π) be a principal G-bundle, V a vector space and $\rho : G \to \mathrm{GL}(V)$ the corresponding linear representation of G and $E := P \times_{(G,\rho)} V$ the associated vector bundle. We want to generalize lemma 2.5 to vector bundle-valued forms.

Definition 2.20. A k-form $\omega \in \Omega^k(P;V)$ is called

- *horizontal* if $\omega_p(X_1,\dots,X_k) = 0$ in case that one of the $X_i \in T_pP$ is vertical;
- *equivariant* if $R_g^*\omega = \rho(g^{-1}) \circ \omega$;
- *basic* if it is both equivariant as well as horizontal.

Remark 2.27. The space of basic k-forms on the total space P is denoted by $\Omega^k_{hor}(P;V)^{(G,\rho)}$.

The next lemma shows that there is a one-to-one correspondence between basic k-forms on P and k-forms on M taking values in the associated bundle E.

Lemma 2.6. *The space $\Omega^k_{hor}(P;V)^{(G,\rho)}$ of basic k-forms on P is isomorphic to the space $\Omega^k(M;E)$ of k-forms on M with values in E.*

Proof. Given a basic k-form $\xi \in \Omega^k_{hor}(P;V)^{(G,\rho)}$, the corresponding k-form $\eta_\xi \in \Omega^k(M;E)$ on the base manifold is defined for each $x \in M$ by

$$\eta_{\xi_x}(X_1,\dots,X_k) := [p, \xi_p(Y_1,\dots,Y_k)],$$

for any $p \in \pi^{-1}(\{x\})$ and $X_j \in T_xM$, $Y_j \in T_pP$ with $\pi_* Y_j = X_j$ for $j = 1,\dots,k$. The definition is independent of the choice of the vectors $Y_j \in T_pP$, since for any other vector $\tilde{Y}_j \in T_pP$ satisfying $\pi_*\tilde{Y}_j = U_j$, we have $\pi_*(Y_j - \tilde{Y}_j) = 0$. It follows that $Y_j - \tilde{Y}_j$ is vertical and as ξ is a horizontal k-form we have $\xi_p(Y_1, \ldots, Y_j - \tilde{Y}_j, \ldots, Y_k) = 0$. Moreover, the definition is also independent of the choice of $p \in \pi^{-1}(\{x\})$ since ξ is an equivariant k-form. Indeed, let $\tilde{Y}_j \in T_{p.g}P$ such that $\pi_*\tilde{Y}_j = U_j$ for all $j = 1, \ldots, k$. Then, we have

$$\begin{aligned}
[p.g, \xi_{p.g}(\tilde{Y}_1, \ldots, \tilde{Y}_k] &= [p.g, \xi_{p.g}(R_{g_*}Y_1, \ldots, R_{g_*}Y_k)] \\
&= [p.g, R_g^*\xi_p(Y_1, \ldots, Y_k)] \\
&= [p.g, \rho(g^{-1})\xi_p(Y_1, \ldots, Y_k)] \\
&= [p, \xi_p(Y_1, \ldots, Y_k)].
\end{aligned}$$

Conversely, let $\eta \in \Omega^k(M; E)$ be a k-form and define $\xi_\eta \in \Omega^k_{hor}(P; V)^{(G,\rho)}$ by

$$\xi_{\eta_p}(Y_1, \ldots, Y_k) \coloneqq \iota_p^{-1}\eta_x(\pi_*Y_1, \ldots, \pi_*Y_k),$$

where $\iota_p : V \to \tilde{\pi}^{-1}(\{x\})$ is the inclusion of the fiber and is defined by $\iota_p(f) \coloneqq [p, f]$ for any $p \in \pi^{-1}(\{x\})$. This so defined k-form ξ_η is basic. Indeed, since for any $v \in V$ we have $\iota_p(v) = \iota_{p.g}(\rho(g^{-1}).v)$ and it follows

$$\begin{aligned}
(R_g^*\xi_\eta)_p(Y_1, \ldots, Y_k) &= \xi_{\eta_{p.g}}(R_{g_*}Y_1, \ldots, R_{g_*}Y_k) \\
&= \iota_{p.g}^{-1}\eta_x((\pi \circ R_g)_*Y_1, \ldots, (\pi \circ R_g)_*Y_k) \\
&= \iota_{p.g}^{-1}\eta_x(\pi_*Y_1, \ldots, \pi_*Y_k) \\
&= \iota_{p.g}^{-1}\iota_p(\xi_{\eta_p}(Y_1, \ldots, Y_k)) \\
&= \rho(g^{-1}).\xi_{\eta_p}(Y_1, \ldots, Y_k)
\end{aligned}$$

where we used that $\pi \circ R_g = \pi$. This shows that ξ_η is equivariant. It is also horizontal, as for any vertical vector $\tilde{Y}_j \in T_pP$ we have $\pi_*\tilde{Y}_j = 0$ □

The Space of Connections

We now use lemma 2.6 to describe the space of connections on a principal bundle.

Corollary 2.1. *Let (P, M, π, G) be a principal bundle. The space of connections $\mathcal{A}(P)$ on P is an affine space modeled over the vector space $\Omega^1(M; \mathrm{ad}(P))$, where $\mathrm{ad}(P) = P \times_{G,\mathrm{ad}} \mathfrak{g}$ is the adjoint bundle.*

2.5 Gauge Transformations

Definition 2.21. Let (P, M, π, G) be a principal bundle. A *gauge transformation* is an automorphism of the principal bundle, i.e. a principal bundle map ϑ making the following diagram commute

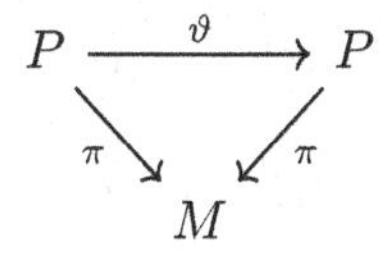

Remark 2.28. The set of all gauge transformations forms a group under composition of automorphisms. We denote the *group of gauge transformations* with $\mathcal{G}(P)$.

Remark 2.29. The group of gauge transformations $\mathcal{G}(P)$ can be identified with the space of sections in the adjoint bundle $\mathrm{Ad}(P) = P \times_{G,\mathrm{Ad}} G$

$$\mathcal{G}(P) \simeq \Gamma(M; \mathrm{Ad}(P)).$$

Indeed, by lemma 2.5 we can identify $\Gamma(M; \mathrm{Ad}(P))$ with the space of equivariant maps $C^\infty(P; G)^{(G,\mathrm{Ad})}$, that is maps γ satisfying $\gamma(p.g) = \mathrm{Ad}(g^{-1})\gamma(p) = g^{-1}\gamma(p)g$. For each such γ we can define a map $\vartheta_\gamma : P \to P$ by

$$\vartheta_\gamma(p) := p.\gamma(p).$$

This is an element of $\mathcal{G}(P)$. Indeed, it is G-equivariant

$$\begin{aligned}\vartheta_\gamma(p.g) &= p.g.\gamma(p.g)\\ &= p.gg^{-1}\gamma(p)g\\ &= \vartheta_\gamma(p).g,\end{aligned}$$

for all $g \in G$ and satisfies $\pi \circ \vartheta_\gamma = \pi$ since $\pi(\vartheta_\gamma(p)) = \pi(p.\gamma(p)) = \pi(p)$. Conversely, given $\vartheta \in \mathcal{G}(P)$, we define $\gamma_\vartheta \in C^\infty(P; G)^{(G,\mathrm{Ad})}$ by $\gamma_\vartheta(p) := \phi(p)^{-1}\phi(\vartheta(p))$, where $\phi : \pi^{-1}(U) \to G$ is a local trivialization.

2.5.1 The Action of the Group of Gauge Transformations on the Space of Connections

There is a natural action of the group of gauge transformations $\mathcal{G}(P)$ on the space of connections $\mathcal{A}(P)$.

Lemma 2.7. *Let $H \subset TP$ be a connection on a principal bundle (P, M, π, G) and let $\vartheta \in \mathcal{G}(P)$ be a gauge transformation. Then, we have that*

$$H^{\vartheta} := \vartheta_* H$$

is also a connection on P.

Lemma 2.8. *Let $\omega \in \Omega^1(P; \mathfrak{g})$ be a connection 1-form on a principal bundle (P, M, π, G) and let $\vartheta \in \mathcal{G}(P)$ be a gauge transformation. Then, we have that*

- *$\vartheta^* \omega$ is also a connection on P;*
- *$\vartheta^* \omega = \mathrm{ad}_{\gamma_{\vartheta}^{-1}} \circ \omega + \gamma_{\vartheta}^* \theta$,*

where θ is the Maurer-Cartan form on G and $\gamma_{\vartheta} \in C^{\infty}(P; G)^{(G, Ad)}$ is the equivariant map associated to $\vartheta \in \mathcal{G}(P)$.

Remark 2.30. Given a connection 1-form ω and the gauge-transformed connection 1-form $\omega^{\vartheta} := \vartheta^{-1^*} \omega$, the corresponding local representatives $A_{\alpha} = s_{\alpha}^* \omega$ and $A_{\alpha}^{\vartheta} = s_{\alpha}^* \omega^{\vartheta}$ are related via

$$A_{\alpha} = \mathrm{ad}_{\bar{\vartheta}_{\alpha}^{-1}} \circ A_{\alpha}^{\vartheta} + \bar{\vartheta}_{\alpha}^* \theta,$$

where $s_{\alpha} : U_{\alpha} \to \pi^{-1}(U_{\alpha})$ is a local section and $\bar{\vartheta}_{\alpha}$ is a function $\bar{\vartheta}_{\alpha} : U_{\alpha} \to G$ defined via

$$\bar{\vartheta}_{\alpha}(\pi(p)) := \phi_{\alpha}(\vartheta(p)) \phi_{\alpha}(p)^{-1},$$

where $\phi_{\alpha} : \pi^{-1}(U_{\alpha}) \to G$ is the local trivialization map canonically associated to the local section. Also notice that if G is a matrix Lie group, we can write

$$A_{\alpha} = \bar{\vartheta}_{\alpha}^{-1} A_{\alpha}^{\vartheta} \bar{\vartheta}_{\alpha} + \bar{\vartheta}_{\alpha} d\bar{\vartheta}_{\alpha}.$$

2.6 The Curvature of a Connection

Throughout, let (P, M, π, G) be a principal bundle equipped with a connection 1-form ω and denote with $\mathrm{ad}(P) = P \times_{G,\mathrm{ad}} G$ the adjoint bundle over M.

Given a connection, we show how to define a *covariant exterior derivative* on the space of differential forms on M with values in the adjoint bundle $\mathrm{ad}(P)$. The obstruction of the covariant exterior derivative to define a complex is then measured by the *curvature* of the connection. We also show the transformation properties of the connection under gauge transformations.

2.6.1 The Covariant Exterior Derivative

Our goal is to define the *covariant exterior derivative*

$$\bar{d}_\omega : \Omega^k(M; \mathrm{ad}(P)) \to \Omega^{k+1}(M; \mathrm{ad}(P)),$$

induced by the connection 1-form ω. By lemma 2.6 we can identify the forms on M with values in the adjoint bundle with horizontal, equivariant $\mathfrak{g}$-valued forms on P. Thus, we are searching for a map that assigns to a basic $\mathfrak{g}$-valued form on P another basic form.

Definition 2.22. Let V be a vector space and denote with $\Omega^\bullet(P; V)$ the space of differential forms on P with values in V. This space is equipped with a linear map $d : \Omega^k(P; V) \to \Omega^{k+1}(P; V)$, $\omega \mapsto d\omega$, defined for $k > 0$ by

$$\begin{aligned} d\omega(X_0, \ldots, X_k) = &\textstyle\sum_{i=0}^k (-1)^i X_i \omega(X_0, \ldots, \not{X}_i, \ldots, X_k) + \\ &\textstyle\sum_{i<j} (-1)^{i+j} \omega([X_i, X_j], X_0, \ldots, \not{X}_i, \ldots, \not{X}_j, \ldots, X_k), \end{aligned}$$

satisfying $d \circ d = 0$ and thus, turning $(\Omega^\bullet(P; V), d)$ into a complex.

Remark 2.31. Let $\xi \in \Omega^k_{hor}(P; \mathfrak{g})^{(G,\mathrm{ad})}$ be a basic form. Then we have that $d\xi$ is equivariant. Indeed, we have

$$\begin{aligned} R_g^*(d\xi) &= d(R_g^*\xi) \\ &= d(\mathrm{ad}_{g^{-1}} \circ \xi) \\ &= \mathrm{ad}_{g^{-1}} \circ d\xi. \end{aligned}$$

However, $d\xi$ need not be horizontal. Thus, the differential d is not the map we are looking for.

The data of a connection allows us to define a projection on horizontal forms, i.e. on forms that vanish on vertical vectors. This *horizontal projection* is then used to define the exterior derivative $d_\omega\xi$ as the horizontal part of $d\xi$.

Definition 2.23. Let $H \subset TP$ be a connection in a principal bundle (P, M, π, G). Define the *horizontal projection*

$$h : TP \to TP$$

as a collection of linear maps $h_p : T_pP \to T_pP$, satisfying

$$h_p(v) := \begin{cases} v, & \text{if } v \in H_pP, \\ 0, & \text{if } v \in V_pP, \end{cases}$$

for all $p \in P$.

Remark 2.32. The horizontal projection induces a dual map $h^* : T^*P \to T^*P$, defined for every $p \in P$ by

$$(h_p^*(f))(v) = f(h_p(v))$$

for $f \in T_p^*P$ and $v \in T_pP$. More generally, we can extend h to a map

$$h_k : \bigwedge^k(TP) \to \bigwedge^k(TP)$$

by setting $h_k(v_1 \wedge \cdots \wedge v_k) = h(v_1) \wedge \cdots \wedge h(v_k)$. For simplicity we write $h_k = h$. Then, for $\omega \in \Omega^k(P)$ we have $h^*\omega(v_1, \dots, v_k) = \omega(h(v_1), \dots, h(v_k))$.

Definition 2.24. Let (P, M, π, G) be a principal bundle equipped with a connection 1-form ω. The *covariant exterior derivative* d_ω is defined as

$$d_\omega : \Omega^k(P; \mathfrak{g}) \to \Omega^{k+1}(P; \mathfrak{g}), \quad \xi \mapsto d\xi \circ h.$$

Lemma 2.9. *The exterior covariant derivative d_ω maps horizontal, equivariant $\mathfrak{g}$-valued k-forms into horizontal, equivariant $\mathfrak{g}$-valued $(k+1)$-forms on P*

$$d_\omega : \Omega^k_{hor}(P; \mathfrak{g})^{(G,\mathrm{ad})} \to \Omega^{k+1}_{hor}(P; \mathfrak{g})^{(G,\mathrm{ad})}.$$

The following lemma gives an explicit formula for the covariant exterior derivative.

Lemma 2.10. *For every k-form $\xi \in \Omega^k_{hor}(P; \mathfrak{g})^{(G,\mathrm{ad})}$ we have*

$$d_\omega \xi = d\xi + [\omega \wedge \xi].$$

Remark 2.33. By lemma 2.6, the exterior covariant derivative d_ω induces an exterior covariant derivative $\bar{d}_\omega$ on the space of differential forms on M with values in the adjoint bundle, we thus have a commutative diagram

$$\begin{array}{ccc} \Omega^k_{hor}(P;\mathfrak{g})^{(G,\mathrm{ad})} & \xrightarrow{d_\omega} & \Omega^{k+1}_{hor}(P;\mathfrak{g})^{(G,\mathrm{ad})} \\ \simeq\downarrow & & \downarrow\simeq \\ \Omega^k(M;\mathrm{ad}(P)) & \xrightarrow{\bar{d}_\omega} & \Omega^{k+1}(M;\mathrm{ad}(P)) \end{array}$$

2.6.2 The Curvature 2-Form

We have characterized the exterior derivative $\bar{d}_\omega$ on $\Omega^\bullet(M;\mathrm{ad}(P))$, however, this is not a differential. Indeed, let $\xi \in \Omega^k_{hor}(P;\mathfrak{g})^{(G,\mathrm{ad})}$ represent a k-form on M with values in the adjoint bundle. Then, we have

$$\begin{aligned} d_\omega d_\omega \xi &= d_\omega(d\xi + [\omega \wedge \xi] \\ &= d^2\xi + d[\omega \wedge \xi] + [\omega \wedge d\xi] + [\omega \wedge [\omega \wedge \xi]] \\ &= [d\omega + \frac{1}{2}[\omega \wedge \omega] \wedge \xi]. \end{aligned}$$

This finding motivates the following definition.

Definition 2.25. Let $\omega \in \Omega^1(P;\mathfrak{g})$ be a connection 1-form on a principal bundle (P, M, π, G). The *curvature 2-from* Ω is defined by

$$\Omega := d_\omega \omega \in \Omega^2(P;\mathfrak{g}).$$

Remark 2.34. Since the connection 1-form ω is equivariant, we have by lemma 2.9 that Ω is a horizontal, equivariant 2-form. Hence, it can be identified with a 2-form $F \in \Omega^2(M;\mathrm{ad}(P))$ defined globally on the base manifold M with values in the adjoint bundle.

Lemma 2.11. *The curvature 2-form Ω satisfies the following* structure equation

$$\Omega = d\omega + \frac{1}{2}[\omega \wedge \omega].$$

Definition 2.26. A connection 1-form on a principal bundle is called *flat* if its curvature is identically zero. A *flat principal bundle* is a principal bundle equipped with a flat connection.

2.6.3 Local Representatives of the Curvature 2-Form and Gauge Transformations

We can pullback the curvature 2-form $\Omega \in \Omega^2(P;\mathfrak{g})$ along local sections $s_\alpha : U_\alpha \to \pi^{-1}(U_\alpha)$ to obtain *local representatives of the curvature 2-form*

$$F_\alpha := s_\alpha^*\Omega \in \Omega^2(U_\alpha;\mathfrak{g}).$$

Notice that in the physics literature the local representatives are often referred to as *gauge field strength.*

Lemma 2.12. *Let $\Omega \in \Omega^2(P;\mathfrak{g})$ be the curvature of a connection 1-form ω on a principal bundle (P, M, π, G) and let $s_\alpha : U_\alpha \to \pi^{-1}(U_\alpha)$ and $s_\beta : U_\beta \to \pi^{-1}(U_\beta)$ be two local sections on $U_\alpha, U_\beta \subset M$. Moreover, let $\phi_{\alpha\beta} : U_\alpha \cap U_\beta \to G$ be the unique transition function such that $s_\beta(x) = s_\alpha(x).\phi_{\alpha\beta}(x)$, for all $x \in U_\alpha \cap U_\beta$. Denote with F_α and F_β the local representatives of Ω with respect to s_α and s_β. Then for $U_\alpha \cap U_\beta \neq 0$ the two local representatives are related by*

$$F_\beta = ad_{\phi_{\alpha\beta}^{-1}} \circ F_\alpha.$$

Remark 2.35. Given a curvature 2-form Ω and a gauge transformation $\vartheta \in \mathcal{G}(P)$, the *gauge-transformed curvature 2-form* is given by

$$\Omega^\vartheta = \vartheta^{-1*}\Omega.$$

Furthermore, one can show that the local representatives $F_\alpha = s_\alpha^*\Omega$ and $F_\alpha^\vartheta = s_\alpha^*\Omega^\vartheta$ transform as

$$F_\alpha^\vartheta = \mathrm{ad}_{\bar{\vartheta}} \circ F_\alpha,$$

where $\bar{\vartheta} : U_\alpha \to G$.

3 Differential Graded Algebras

The purpose of this chapter is to review elementary facts about homological algebra used in the scope of this thesis. In a first part we recall the notions of associative and commutative algebras and establish in this context the terminology for the tensor-, symmetric- and exterior algebra. We also cover Lie algebras and their representations. We then review the basic facts about graded vector spaces and chain complexes and finally, we address differential graded algebras.

Throughout, we use the following notations and conventions. We denote by k a field of characteristic 0. All vector spaces are over k. The category of vector spaces is denoted **Vect**. For V a finite dimensional vector space, we use the notation V^* for $\mathrm{Hom}_{\mathbf{Vect}}(V, k)$. The tensor product of two vector spaces, say V and W, is often denoted by $V \otimes W$ instead of $V \otimes_k W$. The book of J. Loday and B. Vallette [LV12] serves as the main reference for the following. For more details, we also refer to [Wei95], [HS97].

3.1 Algebras

3.1.1 Associative Algebras

Definition 3.1. An *associative algebra* over k is a vector space A endowed with a linear map

$$\mu : A \otimes A \to A,$$

called *product*, such that the following diagram is commutative

$$\begin{array}{ccc} A \otimes A \otimes A & \xrightarrow{\mathrm{id}_A \otimes \mu} & A \otimes A \\ {\scriptstyle \mu \otimes \mathrm{id}_A}\downarrow & & \downarrow{\scriptstyle \mu} \\ A \otimes A & \xrightarrow{\quad \mu \quad} & A \end{array}$$

C. Keller, *Chern-Simons Theory and Equivariant Factorization Algebras*, BestMasters, https://doi.org/10.1007/978-3-658-25338-7_3

Remark 3.1. We often abbreviate the product as $ab := \mu(a \otimes b)$. Then, the associativity condition reads

$$(ab)c = a(bc),$$

for all $a, b, c \in A$.

Definition 3.2. An associative algebra A is said to be *unital* if it has a unit element id_A such that

$$\mathrm{id}_A a = a = a\mathrm{id}_A,$$

for all $a \in A$.

Remark 3.2. In the following, all algebras are unital.

Definition 3.3. Let A and B be algebras over k. A *morphism of algebras* is a linear map $f : A \to B$ such that

$$f(a\tilde{a}) = f(a)f(\tilde{a}) \quad \text{and} \quad f(\mathrm{id}_A) = \mathrm{id}_B$$

for all $a, \tilde{a} \in A$.

Remark 3.3. Associative algebras and the morphisms between them define the category **asAlg**.

As a first example, we review the notion of the tensor algebra of vector spaces. In lemma 3.1 we state the universal property satisfied by the tensor algebra, expressing, roughly speaking, that the tensor algebra is the most general associative algebra containing V.

Example 3.1. Let V be a vector space. The *tensor algebra* $T(V)$ of V is defined as the direct sum of all tensor powers of V, i.e.

$$T(V) = \bigoplus_{n\geq 0} T^n(V) := \bigoplus_{n\geq 0} V^{\otimes n},$$

where by convention we have $T^0(V) = k$. Multiplication in $T(V)$ is given by the tensor product

$$\mu : T^n(V) \otimes T^m(V) \to T^{n+m}(V)$$
$$\mu((v_1 \otimes \cdots \otimes v_n) \otimes (\tilde{v}_1 \otimes \cdots \otimes \tilde{v}_m)) := v_1 \otimes \cdots \otimes v_n \otimes \tilde{v}_1 \otimes \cdots \otimes \tilde{v}_m$$

and is extended by linearity and distributivity to all of $T(V)$.

Remark 3.4. Given a homogeneous element $v \in T^n(V)$, we call the integer n the *weight* of v.

Lemma 3.1. *(Universal property of the tensor algebra) Let A be an associative algebra over k and V a vector space. Given a linear map $f : V \to A$, there exists a unique algebra morphism $\bar{f} : T(V) \to A$ such that the following diagram is commutative*

$$\begin{array}{ccc} V & \xrightarrow{\iota} & T(V) \\ & {\scriptstyle f}\searrow & \downarrow{\scriptstyle \bar{f}} \\ & & A \end{array}$$

where ι is the inclusion of V into $T(V)$.

Remark 3.5. The above lemma shows that the tensor algebra $T(V)$ is *free* in the category of associative algebras. In other words, T is a functor from the category of vector spaces to the category of associative algebras, which is left adjoint to the forgetful functor

$$U : \mathbf{asAlg} \to \mathbf{Vect},$$

assigning to an algebra its underlying vector space. Hence, we have a natural isomorphism

$$\mathrm{Hom}_{\mathbf{asAlg}}(T(V), A) \cong \mathrm{Hom}_{\mathbf{Vect}}(V, U(A)).$$

Example 3.2. Let V be a vector space. The *exterior algebra* $\bigwedge(V)$ of V is defined as

$$\bigwedge(V) := T(V)/I_\wedge,$$

i.e. it is the quotient of the tensor algebra by the two-sided ideal $I_\wedge$ generated by the subset $\{u \otimes v + v \otimes u \mid u, v \in V\}$. The image of $T^n(V)$ in $\bigwedge(V)$ is denoted as $\bigwedge^n(V)$ for all $n \geq 0$ and thus the exterior algebra admits the following direct sum decomposition

$$\bigwedge(V) = \bigoplus_{n \geq 0} \bigwedge^n(V).$$

The induced product $\bigwedge(V) \otimes \bigwedge(V) \to \bigwedge(V)$ is denoted by $\wedge$ and is called the *exterior product* or *wedge product.*

Remark 3.6. We say that the exterior product is *weighted commutative*, that is, for $u \in \bigwedge^n(V)$ and $v \in \bigwedge^m(V)$ we have

$$u \wedge v = (-1)^{nm} v \wedge u.$$

3.1.2 Commutative Algebras

Definition 3.4. A *commutative algebra* over k is an associative algebra A for which the product $\mu : A \otimes A \to A$ is commutative, i.e. we have

$$ab = ba$$

for all $a, b \in A$.

Remark 3.7. In terms of the *switching map*

$$\tau : A \otimes A \to A \otimes A \quad \tau(a \otimes b) := b \otimes a,$$

the commutativity condition reads

$$\mu \circ \tau = \mu.$$

Remark 3.8. Commutative algebras and the morphisms between them define the category **cAlg**. Since all commutative algebras are associative, there is a forgetful functor

$$U : \mathbf{cAlg} \to \mathbf{asAlg}.$$

In analogy to the tensor algebra, we now introduce the free algebra in the category of commutative algebras, namely the symmetric algebra.

Example 3.3. Let V be a vector space. The *symmetric algebra* $\mathrm{Sym}(V)$ of V is defined as

$$\mathrm{Sym}(V) := T(V)/I_S,$$

i.e. it is the quotient of the tensor algebra by the two-sided ideal I_S generated by the subset $\{u \otimes v - v \otimes u \mid u, v \in V\}$. The image of $T^n(V)$ in $\mathrm{Sym}(V)$ is denoted as $\mathrm{Sym}^n(V)$ for all $n \geq 0$ and thus the symmetric algebra admits the following direct sum decomposition

$$\mathrm{Sym}(V) = \bigoplus_{n\geq 0} \mathrm{Sym}^n(V).$$

The induced product $\mathrm{Sym}(V) \otimes \mathrm{Sym}(V) \to \mathrm{Sym}(V)$ is denoted by $\cdot$ and the commutativity condition reads

$$u \cdot v = v \cdot u,$$

for any $u, v \in \mathrm{Sym}(V)$.

Lemma 3.2. *(Universal property of the symmetric algebra) Let B be a commutative algebra over k and V a vector space. Given a linear map $f : V \to B$, there exists a unique algebra morphism $\bar{f} : \mathrm{Sym}(V) \to B$ such that the following diagram is commutative*

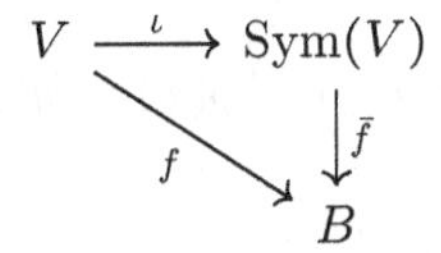

where ι is the inclusion of V into $\mathrm{Sym}(V)$.

Remark 3.9. By the above lemma, we have that Sym is a functor which is left adjoint to the forgetful functor

$$U : \mathbf{cAlg} \to \mathbf{Vect},$$

assigning to a commutative algebra its underlying vector space. Hence, we have a natural isomorphism

$$\mathrm{Hom}_{\mathbf{cAlg}}(\mathrm{Sym}(V), B) \cong \mathrm{Hom}_{\mathbf{Vect}}(V, U(B)).$$

3.1.3 Lie Algebras

We briefly recall the notion of Lie algebras and their representations and introduce in this context the universal enveloping algebra. Further related aspects, such as differential graded Lie algebras and their homotopical generalizations, are treated later in chapter 4 when discussing their role in deformation theory.

Definition 3.5. A *Lie algebra* is a vector space $\mathfrak{g}$ endowed with a linear map

$$[-,-]:\mathfrak{g}\otimes\mathfrak{g}\to\mathfrak{g},$$

called *bracket*, satisfying the following properties

– *(Antisymmetry)*

$$[x,y]=-[y,x];$$

– *(Jacobi identity)*

$$[x,[y,z]]+[y,[z,x]]+[z,[x,y]]=0,$$

for all $x,y,z\in\mathfrak{g}$.

Definition 3.6. Let $(\mathfrak{g},[-,-]_{\mathfrak{g}})$ and $(\mathfrak{h},[-,-]_{\mathfrak{h}})$ be Lie algebras. A *morphism of Lie algebras* is a linear map $\phi:\mathfrak{g}\to\mathfrak{h}$ that is compatible with the bracket, namely it satisfies

$$\phi([x,y]_{\mathfrak{g}})=[\phi(x),\phi(y)]_{\mathfrak{h}},$$

for all $x,y\in\mathfrak{g}$.

Remark 3.10. Lie algebras and the morphisms between them form the category **LieAlg**.

Remark 3.11. Notice that any associative algebra A can be made into a Lie algebra, denoted A_{Lie}, by defining the bracket to be the commutator bracket

$$[a,b]:=ab-ba,$$

for all $a,b\in A$. Hence, we get a forgetful functor

$$()_{\mathrm{Lie}}:\mathbf{asAlg}\to\mathbf{LieAlg}$$

by viewing an associative algebra simply as its underlying vector equipped with the structure of a commutator bracket.

Definition 3.7. Let $\mathfrak{g}$ be a Lie algebra. A $\mathfrak{g}$-*module*, also called a *representation* of $\mathfrak{g}$, is a vector space M equipped with a linear map

$$\rho : \mathfrak{g} \otimes M \to M,$$

such that

$$\rho([x, y] \otimes m) = \rho(x \otimes \rho(y \otimes m)) - \rho(y \otimes \rho(x \otimes m)),$$

for all $x, y \in \mathfrak{g}$ and $m \in M$.

Definition 3.8. Let (M, ρ_M) and (N, ρ_N) be two modules over a Lie algebra $\mathfrak{g}$. A *morphism of* $\mathfrak{g}$*-modules* is a linear map $f : M \to N$, such that

$$f(\rho_M(x \otimes m)) = \rho_N(x \otimes f(m)),$$

for all $x \in \mathfrak{g}$ and $m \in M$.

Remark 3.12. Modules over a Lie algebra $\mathfrak{g}$ and the morphisms between them form the category $\mathfrak{g}$**-Mod** of Lie algebra representations.

Universal Enveloping Algebra

The universal enveloping algebra $U(\mathfrak{g})$ of a Lie algebra $\mathfrak{g}$ is a functor from the category of Lie algebras to the category of associative algebras, which is left adjoint to the forgetful functor $()_{\mathrm{Lie}} : \mathbf{asAlg} \to \mathbf{LieAlg}$. It turns out that $U(\mathfrak{g})$ is the most general associative algebra containing all representations of the Lie algebra $\mathfrak{g}$. As a consequence, we can view $\mathfrak{g}$**-Mod** as the category of left modules over $U(\mathfrak{g})$. This equivalence is used for instance in defining Lie algebra cohomology. The universal enveloping algebra is constructed as follows.

Definition 3.9. Let $\mathfrak{g}$ be a Lie algebra. The *universal enveloping algebra* $U(\mathfrak{g})$ of $\mathfrak{g}$ is defined as

$$U(\mathfrak{g}) := T(\mathfrak{g})/I_U,$$

i.e. it is the quotient of the tensor algebra of $\mathfrak{g}$ by the two-sided ideal I_U generated by the subset $\{x \otimes y - y \otimes x - [x, y] \mid x, y \in \mathfrak{g}\}$

Lemma 3.3. *The functor* $U : \mathbf{LieAlg} \to \mathbf{asAlg}$ *is left adjoint to the forgetful functor* $()_{Lie} : \mathbf{asAlg} \to \mathbf{LieAlg}$. *In other words, we have a natural isomorphism*

$$\mathrm{Hom}_{\mathbf{LieAlg}}(\mathfrak{g}, A_{Lie}) \cong \mathrm{Hom}_{\mathbf{asAlg}}(U(\mathfrak{g}), A).$$

Remark 3.13. The above lemma implies that the category $\mathfrak{g}$-**Mod** is equivalent to the category $U(\mathfrak{g})$-**Mod** of left $U(\mathfrak{g})$ modules. Indeed, let V be a vector space and denote with $\mathrm{End}(V)$ the associative algebra of vector space endomorphisms of V. Then, by lemma 3.3 we have

$$\mathrm{Hom}_{\mathbf{LieAlg}}(\mathfrak{g}, \mathfrak{gl}(V)) \cong \mathrm{Hom}_{\mathbf{asAlg}}(U(\mathfrak{g}), \mathrm{End}(V)),$$

where $\mathfrak{gl}(V)$ is the Lie algebra of endomorphisms of V.

3.2 Graded Vector Spaces

Definition 3.10. A $\mathbb{Z}$-*graded vector space* V is a family of vector spaces $\{V_n\}_{n\in\mathbb{Z}}$.

Remark 3.14. Every $\mathbb{Z}$-graded vector space V admits a direct sum decomposition, which is denoted by

$$V_\bullet := \bigoplus_{n\in\mathbb{Z}} V_n.$$

By abuse of notation we often write V instead of $V_\bullet$.

Remark 3.15. The elements $v \in V_n$ are called *homogeneous of degree* n and we denote the degree by $|v| = n$.

Remark 3.16. It is useful to introduce the following notation. We write $V^n := V_{-n}$ and we refer to $V^\bullet = \bigoplus_n V^n$ as *cohomologically graded* and to $V_\bullet = \bigoplus_n V_n$ as *homologically graded.*

Definition 3.11. Let V and W be two graded vector spaces. A *morphism of graded vector spaces* $f : V \to W$ is a family of degree preserving linear maps $\{f_n : V_n \to W_n\}_{n\in\mathbb{Z}}$.

Remark 3.17. Graded vector spaces and the morphisms between them define the category **gVect**.

Notice that the category **gVect** is enriched over itself if we consider *non-degree preserving* linear maps between the family of vector spaces with grading induced by degree, more precisely we have the following definition.

Definition 3.12. Let V and W be two graded vector spaces. A linear map $f : V \to W$ *of degree* r is a family of linear maps $\{f_n : V_n \to W_{n+r}\}_{n\in\mathbb{Z}}$.

The set of *all* linear maps from V to W is again an element in **gVect**, called the *graded* Hom *space*, and we denote it by

$$\underline{\mathrm{Hom}}(V,W) \coloneqq \bigoplus_{r\in\mathbb{Z}} \underline{\mathrm{Hom}}_r(V,W),$$

where $\underline{\mathrm{Hom}}_r(V,W)$ is the set of linear maps of degree r.

Remark 3.18. A morphism $V \to W$ in **gVect** corresponds to an element in $\underline{\mathrm{Hom}}_0(V,W)$.

Definition 3.13. Let V be a graded vector space and let $r \in \mathbb{Z}$. The *r-suspension* of V is given by the *shift functor*

$$\begin{aligned}[r] : \mathbf{gVect} &\to \mathbf{gVect}\\ V &\to V[r],\end{aligned}$$

where $V[r]_n \coloneqq V_{n-r}$.

Remark 3.19. Notice that a linear map $V \to W$ of degree r can be regarded as a morphism of graded vector spaces $V[r] \to W$.

Definition 3.14. Let V be a graded vector space. We define its *graded dual* V^* to be the graded vector space

$$V^* \coloneqq \underline{\mathrm{Hom}}(V,k).$$

Remark 3.20. Notice that the field k can be regarded as a graded vector space concentrated in degree 0. Hence, we have

$$\begin{aligned}(V^*)_r &= \underline{\mathrm{Hom}}_r(V,k)\\ &= \mathrm{Hom}_{\mathbf{gVect}}(V[r],k)\\ &= \mathrm{Hom}_{\mathbf{Vect}}(V_{-r},k)\\ &= (V_{-r})^*\end{aligned}$$

Definition 3.15. Let V and W be two graded vector spaces. The *tensor product* $V \otimes W$ is defined to be the graded vector space with grading

$$(V\otimes W)_n \coloneqq \bigoplus_{i+j=n} V_i \otimes W_j.$$

Koszul Sign Convention

The monoidal category $(\mathbf{gVect}, \otimes)$ is equipped with a symmetric structure

$$\tau_{V,W} : V \otimes W \to W \otimes V,$$

called the *braiding map.*

Definition 3.16. We call $(\mathbf{gVect}, \otimes, \tau)$ the category of *sign-graded* vector spaces, if the symmetric structure is defined by the *Koszul braiding*

$$\tau_{V,W}(v \otimes w) \coloneqq (-1)^{|v||w|} w \otimes v,$$

for homogeneous elements $v \in V$ and $w \in W$.

Remark 3.21. Notice that we can equip $(\mathbf{gVect}, \otimes, \tau)$ with another choice of symmetry by defining the braiding map as

$$\tau(v \otimes w) \coloneqq w \otimes v,$$

for homogeneous elements $v \in V$ and $w \in W$.

Remark 3.22. Throughout, we mean by **gVect** the category of sign-graded vector spaces.

When working in **gVect**, the sign conventions can be worked out by applying the *Koszul sign rule*, that is when two adjacent symbols, say x and y, switch in a formula, a sign of $(-1)^{|x||y|}$ is required.

Example 3.4. Let $\varphi \in \underline{\mathrm{Hom}}(V, \tilde{V})$ and $\phi \in \underline{\mathrm{Hom}}(W, \tilde{W})$. Then, the map coming from the tensor product of linear maps between graded vector spaces

$$(\varphi \otimes \phi) : V \otimes W \to \tilde{V} \otimes \tilde{W},$$

is defined according to the Koszul sign convention by

$$(\varphi \otimes \phi)(v \otimes w) \coloneqq (-1)^{|\phi||v|} \varphi(v) \otimes \phi(w).$$

3.3 Chain Complexes

Definition 3.17. A *differential graded vector space* (V, d) (dg vector space), also called a *chain complex*, is a graded vector space $V_\bullet$ equipped with a linear map $d : V \to V$ of degree -1, called the *differential* or *boundary map*, satisfying $d_{n-1} \circ d_n = 0$.

$$\dots \xleftarrow{d_{-1}} V_{-1} \xleftarrow{d_0} V_0 \xleftarrow{d_1} V_1 \xleftarrow{d_2} \dots \xleftarrow{d_{n-1}} V_{n-1} \xleftarrow{d_n} V_n \xleftarrow{d_{n+1}} \dots$$

Similarly, a *cochain complex* (V, d) is a graded vector space $V^\bullet$ equipped with a linear map $d : V \to V$ of degree 1, called the *differential* or *coboundary map*, satisfying $d_{n+1} \circ d_n = 0$.

$$\dots \xrightarrow{d_{-2}} V^{-1} \xrightarrow{d_{-1}} V^0 \xrightarrow{d_0} V^1 \xrightarrow{d_1} \dots \xrightarrow{d_{n-2}} V^{n-1} \xrightarrow{d_{n-1}} Vn \xrightarrow{d_n} \dots$$

In the following, we establish the main terminology and notions for chain complexes. A very similar discussion can be done for cochain complexes.

Definition 3.18. Let (V, d_V) and (W, d_W) be two chain complexes. A *morphism of chain complexes* $f : (V, d_V) \to (W, d_W)$, also called a *chain map*, is a morphism of graded vector spaces commuting with the differential

$$d_W \circ f = f \circ d_V.$$

Remark 3.23. Pictorially, a morphism $f : (V, d_V) \to (W, d_W)$ of chain complexes is a family of linear maps $\{f_n : V_n \to W_n\}_{n \in \mathbb{Z}}$ such that the following diagram is commutative

$$\begin{array}{ccc} V_n & \xrightarrow{d_V} & V_{n-1} \\ {\scriptstyle f_n}\downarrow & & \downarrow{\scriptstyle f_{n-1}} \\ W_n & \xrightarrow{d_W} & W_{n-1} \end{array}$$

Remark 3.24. Chain complexes (dg vector spaces) and the morphisms between them define the category **dgVect**.

Notice that the inner Hom space $\underline{\mathrm{Hom}}(V, W)$ has the structure of a chain complex with differential given by the *derivative* of linear maps, defined as follows.

Definition 3.19. Let (V, d_V) and (W, d_W) be two chain complexes. The *derivative* ∂ of a linear map $f \in \underline{\mathrm{Hom}}(V, W)$ is defined for homogeneous f by

$$\partial(f) := d_W \circ f - (-1)^{|f|} f \circ d_V.$$

Definition 3.20. Let $(\underline{\mathrm{Hom}}(V,W),\partial)$ be a chain complex. We call a homogeneous element $f \in \underline{\mathrm{Hom}}(V,W)$ *compatible* with differentials if $\partial f = 0$, that is if we have

$$d_W \circ f = (-1)^{|f|} f \circ d_V.$$

Remark 3.25. A morphism $(V,d_V) \to (W,d_W)$ in **dgVect** corresponds to an element in $f \in \underline{\mathrm{Hom}}_0(V,W)$ that is compatible with differentials.

Definition 3.21. Let (V,d) be a chain complex and let $r \in \mathbb{Z}$. The r-*suspension* of (V,d) is the chain complex

$$(V[r],d[r]), \quad \text{where} \quad V[r]_n \coloneqq V_{n-r} \text{ and } d[r]_n \coloneqq (-1)^r d_{n-r}$$

Definition 3.22. Let (V,d) be a chain complex. We define the *dual complex* as the graded dual V^* equipped with the differential d^* of degree 1 defined by

$$d^* f \coloneqq -(-1)^{|f|} f \circ d$$

for homogeneous $f \in V^*$.

Definition 3.23. Let (V,d_V) and (W,d_W) be two chain complexes. The *tensor product chain complex* $(V,d_V) \otimes (W,d_W)$ is the graded vector space $(V \otimes W)_n \coloneqq \bigoplus_{i+j=n} V_i \otimes W_j$ equipped with differential

$$d_{V\otimes W}(v \otimes w) \coloneqq d_V(v) \otimes w + (-1)^{|v|} v \otimes d_W(w).$$

for homogeneous elements $v \in V$ and $w \in W$.

Remark 3.26. We work with the symmetric monoidal category $(\mathbf{dgVect}, \otimes, \tau)$ with the symmetric structure given by the Koszul braiding inherited from **gVect**.

Definition 3.24. A *double complex* is a bigraded vector space $V = \{V_{p,q}\}_{p,q\in\mathbb{Z}}$ equipped with a *vertical differential* $d^v : V_{p,q} \to V_{p-1,q}$ and a *horizontal differential* $d^h : V_{p,q} \to V_{p,q-1}$, such that the following diagram is commutative

$$\begin{array}{ccccccc} & & \vdots & & \vdots & & \\ & & \downarrow{\scriptstyle d^v} & & \downarrow{\scriptstyle d^v} & & \\ \cdots & \xrightarrow{d^h} & V_{p,q} & \xrightarrow{d^h} & V_{p,q-1} & \xrightarrow{d^h} & \cdots \\ & & \downarrow{\scriptstyle d^v} & & \downarrow{\scriptstyle d^v} & & \\ \cdots & \xrightarrow{d^h} & V_{p-1,q} & \xrightarrow{d^h} & V_{p-1,q-1} & \xrightarrow{d^h} & \cdots \\ & & \downarrow{\scriptstyle d^v} & & \downarrow{\scriptstyle d^v} & & \\ & & \vdots & & \vdots & & \end{array}$$

i.e. such that

$$d^h \circ d^v = d^v \circ d^h.$$

Definition 3.25. The *total complex* of a double complex (V, d^v, d^h) is defined by

$$\mathrm{Tot}(V)_n := \bigoplus_{p+q=n} V_{p,q}.$$

Denote with $\omega_{p,q}$ the projection of $\omega \in \mathrm{Tot}(V)_n$ on $V_{p,q}$. The *total differential* $d^{\mathrm{tot}} : \mathrm{Tot}(V)_n \to \mathrm{Tot}(V)_{n-1}$ is defined by

$$(d^{\mathrm{tot}}\omega)_{p,q} := d^v\omega_{p+1,q} + (-1)^p d^h \omega_{p,q+1}.$$

for $\omega \in \mathrm{Tot}(V)_n$ and for all $p, q \in \mathbb{Z}$ satisfying $p + q = n - 1$.

3.3.1 Homology

Definition 3.26. Given a chain complex $(V_\bullet, d)$, its *homology* is defined to be the complex $(H_\bullet(V_\bullet, d), 0)$, where

$$H_n(V, d) := \mathrm{Ker}(d : V_n \to V_{n-1})/\mathrm{Im}(d : V_{n+1} \to V_n).$$

The kernel of d is referred to as *cycle* and the image of d as *boundary*. Similarly, given a cochain complex $(V^\bullet, d)$, the kernel of d is referred to as *cocycle* and the image of d as *coboundary*. Its *cohomology* is defined to be

$$H^n(V, d) := \mathrm{Ker}(d : V^n \to V^{n+1})/\mathrm{Im}(d : V^{n-1} \to V^n).$$

In the following we mainly focus on chain complexes and their homology, but similar notions can be reworked for cochain complexes and cohomology respectively.

Lemma 3.4. *Let $f : (V, d_V) \to (W, d_W)$ be a morphism of chain complexes. We have that f takes cycles to cycles and boundaries to boundaries. Consequently, f descends to a well-defined morphism in homology*

$$f_* \coloneqq H_\bullet(f) : H_\bullet(V, d_V) \to H_\bullet(W, d_W).$$

Remark 3.27. For any $n \in \mathbb{Z}$, we associate with the chain complex (V, d_V) its nth *homology group* $H_n(V, d_V)$ and we associate with every morphism of chain complexed $f : (V, d_V) \to (W, d_W)$ the induced morphism in homology $f_* : H_n(V, d_V) \to H_n(W, d_W)$. If $g : (U, d_U) \to (V, d_V)$ is another morphism of chain complexes, we have

$$(f \circ g)_* = f_* \circ g_*$$

and since for the identity morphism $\mathrm{id} : (V, d_V) \to (V, d_V)$ we have

$$\mathrm{id}_* = \mathrm{id},$$

H_n is a functor, called the nth *homology functor.*

Remark 3.28. Let (V, d) be a chain complex and consider its r-suspension $(V[r]), d[r])$. Then, its homology is given by

$$H_n(V[r], d[r]) = H_{n-r}(V, d).$$

Exact Sequences

Definition 3.27. Let $\{V_n\}_{n\in\mathbb{Z}}$ be a sequence of vector spaces. Then, the sequence of linear maps

$$\cdots \to V_{n+1} \xrightarrow{f_{n+1}} V_n \xrightarrow{f_n} V_{n-1} \to \ldots$$

is said to be *exact* if $\mathrm{Ker}(f_n) = \mathrm{Im}(f_{n+1})$ for all $n \in \mathbb{Z}$.

A *short exact sequence* is an exact sequence of the form

$$0 \to U \xrightarrow{f} V \xrightarrow{g} U \to 0.$$

Definition 3.28. A chain complex (V, d) is called *acyclic* if each segment

$$V_{n+1} \xrightarrow{d} V_n \xrightarrow{d} V_{n-1}$$

is exact.

Remark 3.29. Notice that a chain complex is acyclic if its homology is 0 everywhere.

Lemma 3.5. *Let $f : (V, d_V) \to (W, d_W)$ and $g : (U, d_U) \to (V, d_V)$ be morphisms of chain complexes. Suppose that*

$$0 \to (U, d_U) \xrightarrow{g} (V, d_V) \xrightarrow{f} (W, d_W) \to 0$$

is short exact. Then, there is a long exact sequence in homology of the form

$$\begin{array}{ccccc} \ldots \longrightarrow H_{n+1}(U, d_U) & & & & \\ \downarrow \partial_{n+1} & & & & \\ H_n(U, d_U) & \xrightarrow{g_*} & H_n(V, d_V) & \xrightarrow{f_*} & H_n(W, d_W) \\ & & & & \downarrow \partial_n \\ & & & & H_{n-1}(U, d_U) \longrightarrow \ldots \end{array}$$

where ∂ is referred to as connecting morphism.

Chain Homotopies and Quasi-Isomorphisms

Definition 3.29. Let (V, d_V) and (W, d_W) be two chain complexes and let $f, g \in \underline{\mathrm{Hom}}_k(V, W)$ be two linear maps compatible with differentials. A *chain homotopy* between f and g is an element $h \in \underline{\mathrm{Hom}}_{k+1}(V, W)$ such that

$$\partial h = d_W \circ h + (-1)^k h \circ d_V = f - g.$$

Remark 3.30. We say that f and g are *homotopic* if there exists a homotopy between them and we write $f \simeq g$. In particular, we say that f is *null homotopic* if $f \simeq 0$.

Remark 3.31. The notion of a chain homotopy can also be given for two linear maps $f, g \in \underline{\mathrm{Hom}}_k(V, W)$ that are *not* compatible with differentials, but satisfying the weaker condition

$$\partial f = \partial g.$$

Lemma 3.6. *Given two homotopic chain maps $f, g : (V, d_V) \to (W, d_W)$, the induced morphisms on homology $f_*, g_* : H(V, d_V) \to H(W, d_W)$ are equal.*

Definition 3.30. Let (V, d_V) and (W, d_W) be two chain complexes and let $f \in \underline{\mathrm{Hom}}_k(V, W)$ be a linear map compatible with differentials. We call f a *chain homotopy equivalence* if there exists a map $g \in \underline{\mathrm{Hom}}_{-k}(W, V)$ compatible with differentials such that $g \circ f \simeq \mathrm{id}_V$ and $f \circ g \simeq \mathrm{id}_W$.

If $f \in \underline{\mathrm{Hom}}_k(V, W)$ is a chain homotopy equivalence we call the chain complexes (V, d_V) and (W, d_W) *homotopy equivalent.*

Remark 3.32. Pictorially we have

$$h \circlearrowright V \underset{g}{\overset{f}{\rightleftarrows}} W \circlearrowleft j$$

with

$$\mathrm{id}_V - g \circ f = d_V \circ h - (-1)^{|h|} h \circ d_V, \quad \mathrm{id}_W - f \circ g = d_W \circ j - (-1)^{|j|} j \circ d_W.$$

Remark 3.33. Since $g \circ f \simeq \mathrm{id}_V$ and $f \circ g \simeq \mathrm{id}_W$ we have

$$(g \circ f)_* = g_* \circ f_* = \mathrm{id}_V, \quad (f \circ g)_* = f_* \circ g_* = \mathrm{id}_W,$$

that is f_* and g_* are mutually inverse chain isomorphisms.

Definition 3.31. A morphism $f : (V, d_V) \to (W, d_W)$ of chain complexes is called a *quasi-isomorphism* if the induced morphism on homology $f_* : H(V, d_V) \to H(W, d_W)$ is an isomorphism.

Remark 3.34. Notice that every chain homotopy equivalence is a quasi-isomorphism but the converse is in general not true.

3.4 Graded Algebras

Definition 3.32. An associative *graded algebra* A is a graded vector space $\{A_n\}_{n\in\mathbb{Z}}$ endowed with an associative product $\mu : A \otimes A \to A$ respecting the grading

$$\mu : A_p \otimes A_q \to A_{p+q},$$

for all $p, q \in \mathbb{Z}$.

Remark 3.35. Equivalently, an associative graded algebra is a monoid object in $(\mathbf{gVect}, \otimes)$, that is an object A of **gVect** endowed with a multiplication morphism $\gamma : A \otimes A \to A$ and a unit morphism $\eta : k \to A$ that satisfy the associativity and unity axioms.

Remark 3.36. Let A be a graded algebra. Notice that the r-suspension $A[r]$ is *not* a graded algebra, since the product in $A[r]$ does not respect the grading. Indeed, for $a \in (A[r])_p$ and $b \in (A[r])_q$, the product ab has degree $|ab| = |a| + |b| - 2r$ and hence it is not an element of $(A[r])_{p+q}$. This is a consequence of the fact that the shift functor is not monoidal.

Definition 3.33. Let A and B be two graded algebras. A *morphism of graded algebras* $f : A \to B$ is a morphism of the underlying graded vector spaces satisfying

$$f(a\tilde{a}) = f(a)f(\tilde{a}) \quad \text{and} \quad f(\mathrm{id}_A) = \mathrm{id}_B$$

for all $a, \tilde{a} \in A$.

Remark 3.37. Graded algebras and the morphisms between them define the category **gAlg**.

Definition 3.34. A map of graded algebras $f : A \to B$ *of degree* r is an element $f \in \underline{\mathrm{Hom}}_r(A, B)$ that is compatible with multiplication and preserves the unit element.

Definition 3.35. Let A and B be two graded algebras. The *tensor product graded algebra* is the graded vector space $(A \otimes B)_n = \bigoplus_{i+j=n} A_i \otimes B_j$ together with the following product, defined for all homogeneous elements $a, \tilde{a} \in A$ and $b, \tilde{b} \in B$ by

$$(a \otimes b)(\tilde{a} \otimes \tilde{b}) := (-1)^{|\tilde{a}||b|} a\tilde{a} \otimes b\tilde{b}.$$

Definition 3.36. A *commutative graded algebra* is a graded algebra A for which the product $\mu : A \otimes A \to A$ is graded commutative, i.e. we have for all homogeneous elements $a, b \in A$

$$ab = (-1)^{|a||b|} ba.$$

Example 3.5. Let A and B be two graded commutative algebras. The tensor product of A and B is again a graded commutative algebra. Indeed, we have

$$\begin{aligned}(a \otimes b)(\tilde{a} \otimes \tilde{b}) &= (-1)^{|\tilde{a}||b|} a\tilde{a} \otimes b\tilde{b} \\ &= (-1)^{|\tilde{a}||b| + |\tilde{a}||a| + |\tilde{b}||b|} \tilde{a}a \otimes \tilde{b}b \\ &= (-1)^{(|a|+|b|)(|\tilde{a}|+|\tilde{b}|)} (\tilde{a} \otimes \tilde{b})(a \otimes b).\end{aligned}$$

Weight Graded Algebras

Consider the tensor algebra $T(V)$ of a graded vector space $\{V_n\}_{n \in \mathbb{Z}}$. Recall from remark 3.4 that we assign to all homogeneous elements in $T^n(V)$ the *weight* n, hence, the tensor algebra $T(V)$ is a direct sum of subspaces $T^n(V)$ indexed by their weight. With V being a graded vector space itself, an element $v = v_1 \otimes \cdots \otimes v_n \in T^n(V)$ has assigned a homological grading, i.e. its degree $|v| = |v_1| + \cdots + |v_n|$, and a non-homological grading, its weight n.

Throughout, we will always ignore the weight. For instance, in the case of the tensor algebra, we define the grading by

$$(T(V))_n \coloneqq \bigoplus_{(p)} T(V_0) \otimes T^{p_1}(V_1) \otimes \cdots \otimes T^{p_r}(V_r),$$

where the direct sum is over all r-tuples $(p_1, \ldots, p_r)$, where the p_i constitute a partition of n, that is they satisfy $\sum_{i \in \mathbb{Z}} p_i i = n$, and the tensor algebra is then given as the direct sum $T(V) \coloneqq \bigoplus_{n \in \mathbb{Z}} (T(V))_n$.

3.5 Differential Graded Algebras

Definition 3.37. Let A be a graded algebra. A linear map $d : A \to A$ of degree p is called a *derivation of degree* p if it satisfies

– (*Graded Leibniz rule*)

$$d(ab) = (da)b + (-1)^{p|a|}adb,$$

for homogeneous elements $a, b \in A$.

Definition 3.38. A *differential graded algebra* (dg algebra) is a graded algebra A equipped with a derivation $d : A \to A$ of degree -1, called the *differential*, satisfying $d^2 = 0$.

Definition 3.39. Let (A, d_A) and (B, d_B) be two dg algebras. A *morphism of dg algebras* $f : (A, d_A) \to (B, d_B)$ is a morphism of graded algebras compatible with the differentials

$$d_B \circ f = f \circ d_A.$$

Remark 3.38. Dg algebras and the morphisms between them define the category **dgAlg**.

Definition 3.40. A map of dg algebras $f : (A, d_A) \to (B, d_B)$ *of degree* r is a map of graded algebras of degree r and we say that it is compatible with the differentials if it satisfies

$$d_B \circ f = (-1)^{|f|} f \circ d_A.$$

Definition 3.41. Let (A, d_A) and (B, d_B) be two dg algebras. The *tensor product dg algebra* is the tensor product graded algebra $A \otimes B$ with differential

$$d_{A\otimes B}(a \otimes b) \coloneqq d_A(a) \otimes b + (-1)^{|a|} a \otimes d_B(b).$$

for homogeneous elements $a \in A$ and $b \in B$.

Definition 3.42. A *commutative dg algebra* (cdg algebra) A is an associative dg algebra equipped with a graded commutative product.

Remark 3.39. Cdg algebras and the morphisms between them define the category **cdgAlg**.

Example 3.6. Let (V, d) be a dg vector space. Then, the *symmetric algebra* $\mathrm{Sym}(V)$ is a cdg algebra. It is obtained form the tensor algebra $T(V) \coloneqq \bigoplus_{n\in\mathbb{Z}} (T(V))_n$ as the quotient

$$\mathrm{Sym}(V) \coloneqq T(V)/I_S,$$

where I_S is the two-sided ideal generated by the subset $\{u \otimes v - (-1)^{|u||v|} v \otimes u \mid u, v \in V\}$.

3.5.1 Homotopies

We introduce the appropriate notion of homotopies in the setting of dg algebras. Recall that for $f, g \in \underline{\mathrm{Hom}}^k(V, W)$ a chain homotopy h between f and g is an element $h \in \underline{\mathrm{Hom}}^{k-1}(V, W)$ satisfying

$$\partial h = f - g.$$

Equivalently, we can regard homotopies between f and g as elements in $\underline{\mathrm{Hom}}^k(V, W \otimes k[t, dt])$ that restrict to f and g for $t = 0$ and $t = 1$ respectively. Here $k[t, dt]$ is a cdg algebra with t in degree 0. Indeed, we can define

$$H(v) := f(v)t + g(v)(1-t) - (-1)^{|v|} h(v)dt$$

for $t \in [0,1]$ and a homogeneous element $v \in V$. H is a map of cochain complexes. Indeed, we have

$$\begin{aligned}
d(f(v)t + g(v)(1-t) - (-1)^{|v|} h(v)dt) \\
&= d(f(v))t + (-1)^{k+|v|} f(v)dt + d(g(v))(1-t) \\
&\quad - (-1)^{k+|v|} g(v)dt - (-1)^{|v|} d(h(v))dt \\
&= (-1)^k f(dv) + (-1)^k g(dv)(1-t) + (-1)^{k+|v|} h(dv)dt \\
&= (-1)^k (f(dv) + g(dv)(1-t) - (-1)^{|dv|} h(dv)dt),
\end{aligned}$$

where we used that $f - g = d \circ h + (-1)^k h \circ d$. These observations motivate the following definition.

Definition 3.43. Let A and B be dg algebras and let $f, g : A \to B$ be a morphism of dg algebras. A *homotopy* between f and g is a morphism of dg algebras

$$H : A \to B \otimes k[t, dt],$$

such that $H|_{t=0} = f$ and $H|_{t=1} = g$.

We then have the following useful lemma.

Lemma 3.7. *Let $f, g : V \to \mathrm{Sym}(W)$ be morphisms of chain complexes and let h be a chain homotopy between f and g. Then h induces a canonical homotopy between $\mathrm{Sym}(f)$ and $\mathrm{Sym}(g)$.*

Proof. Consider the morphism of chain complexes $H : V \to \mathrm{Sym}(W) \otimes k[t, dt]$ associated to h. Since $\mathrm{Sym}(W) \otimes k[t, dt]$ is a commutative dg algebra, it follows from the universal property of the symmetric algebra that we obtain a morphism of algebras

$$\mathrm{Sym}(H) : \mathrm{Sym}(V) \to \mathrm{Sym}(W) \otimes k[t, dt].$$

If we set $t = 0$ and $t = 1$ we obtain $\mathrm{Sym}(f)$ and $\mathrm{Sym}(g)$, respectively. $\square$

4 L_∞-Algebras and Derived Formal Moduli Problems

Broadly speaking, deformation theory deals with families of structures that arise when varying a given object in dependence of some suitable parameter space, comprising the study of *moduli spaces*, which are spaces parameterizing equivalence classes of structures. With a *formal moduli problem* we thus mean the infinitesimal description of a moduli space, capturing the local structure around a given point. In this chapter we first address the classical theory of algebraic deformation problems, before explaining how formal moduli problems arise as deformation functors in algebraic geometry. The functorial perspective allows a natural generalization to derived algebraic geometry. There is a common 'guiding principle' behind all deformation problems, namely that they are governed by certain elements of a dg Lie algebra. The precise statement due to J. Lurie [Lur11] is given in the last section of this chapter.

4.1 Classical Deformation Theory of Associative Algebras

We start this chapter by recalling the classical deformation theory for associative algebras and the cohomological aspects controlling the deformations. This exemplary problem should give some intuition before we turn to a more formal treatment. This section follows closely the lecture notes by M. Markl [MDZ07] on deformation theory.

4.1.1 Deformations of Associative Algebras

We review some definitions given in [MDZ07] for describing *formal deformations* of associative algebras. Throughout, let k be a field of characteristic 0

C. Keller, *Chern-Simons Theory and Equivariant Factorization Algebras*, BestMasters, https://doi.org/10.1007/978-3-658-25338-7_4

and R an *augmented* unital commutative k-algebra. Recall that a k-algebra R is augmented if it is equipped with a morphism of algebras $\epsilon : R \to k$.

Remark 4.1. Notice that k is an R-bimodule with the bimodule structure induced by the augmentation map $\epsilon : R \to k$.

Definition 4.1. Let A be an associative k-algebra. An *R-deformation* of A is an associative R-algebra B together with a k-algebra isomorphism

$$\alpha : \bar{B} := k \otimes_R B \to A.$$

Definition 4.2. Two R-deformations $(B, \alpha : \bar{B} \to A)$ and $(B', \alpha' : \bar{B}' \to A)$ of A are said to be *equivalent* if there exists an R-algebra isomorphism $\phi : B \to B'$ such that

$$\bar{\phi} : \bar{B} \to \bar{B}', \quad \bar{\phi} = \alpha'^{-1} \circ \alpha.$$

In the following, A is always an associative k-algebra and B an associative R-algebra. Moreover, we consider deformations $(B, \underline{\text{can}} : \bar{B} \to A)$, where B is a *free R-module.* Equivalently, we assume that we have an isomorphism of R-modules

$$B \simeq R \otimes A.$$

Then, the canonical map $\underline{\text{can}} : k \otimes_R (R \otimes A) \to A$ is given by

$$\underline{\text{can}}(1 \otimes_R (1 \otimes a)) := a,$$

for $a \in A$.

Remark 4.2. The R-algebra B is equipped with an associative R-bilinear multiplication

$$\mu : B \times B \to B.$$

We assume that each element of $B \simeq R \otimes A$ is a finite sum of elements ra, $r \in R$ and $a \in A$. By R-bilinearity of μ we have

$$\mu(ra, r'a') = rr'\mu(a, a'),$$

for each $r, r' \in R$ and $a, a' \in A$. Hence, multiplication in B is determined by its restriction to $A \otimes A$.

Definition 4.3. A *(one-parameter) formal deformation* is a deformation as given in definition 4.1 over the ring of formal power series $k[[t]]$ in the variable t.

Lemma 4.1. *A formal deformation B of A is given by a family of k-linear maps*

$$\{\mu_i : A \otimes A \to A \mid i \in \mathbb{N}\},$$

satisfying $\mu_0(a,b) = ab$, where μ_0 is the multiplication in A, and

$$\sum_{i+j=m} \mu_i(\mu_j(a,b),c) = \sum_{i+j=m} \mu_i(a,\mu_j(b,c)), \qquad (\star)$$

for all $a,b,c \in A$ and $m \geq 1$.

Proof. Let $\mu : B \otimes_{k[[t]]} B \to B$ denote the multiplication in B. By remark 4.2 we see that the multiplication μ is determined by its restriction to $A \otimes A$. We can expand μ into a formal power series

$$\mu(a,b) = \mu_0(a,b) + t\mu_1(a,b) + t^2\mu_2(a,b) + \dots,$$

for k-linear functions $\mu_i : A \otimes A \to A$, $i \geq 0$. Associativity then reads $\mu(\mu(a,b),c) = \mu(a,\mu(b,c))$ for $a,b,c \in A$. □

4.1.2 Cohomological Aspects

Hochschild Cohomology

Let A be an associative k-algebra and let M be an A-module. We recall the definition of the Hochschild complex and its cohomology.

Definition 4.4. The *Hochschild cohomology* of A with coefficients in M is the cohomology of the complex

$$0 \to M \xrightarrow{\delta} C^1_{\mathrm{Hoch}}(A,M) \xrightarrow{\delta} \dots \xrightarrow{\delta} C^n_{\mathrm{Hoch}}(A,M) \xrightarrow{\delta} \dots,$$

where the cochains $C^n_{\mathrm{Hoch}}(A,M)$ are given by the space of k-linear maps from $A^{\otimes n}$ to M and the differential $\delta : C^n_{\mathrm{Hoch}}(A,M) \to C^{n+1}_{\mathrm{Hoch}}(A,M)$ is defined for all $a_i \in A$ by

$$\begin{aligned}\delta f(a_0 \otimes \dots \otimes a_n) :=& (-1)^{n+1} a_0 f(a_1 \otimes \dots \otimes a_n) + f(a_0 \otimes \dots \otimes a_{n-1})a_n \\ &+ \sum_{i=0}^{n-1} (-1)^{i+n} f(a_0 \otimes \dots \otimes a_i a_{i+1} \otimes \dots \otimes a_n).\end{aligned}$$

Remark 4.3. We usually denote the nth Hochschild cohomology group by

$$H^n_{\text{Hoch}}(A, M) := H^n(C^\bullet_{\text{Hoch}}(A, M), \delta).$$

Infinitesimal Deformations

Remark 4.4. Notice that equation $(\star)$ for $m = 1$ reads

$$a\mu_1(b, c) - \mu_1(ab, c) + \mu_1(a, bc) - \mu_1(a, b)c = 0 = \delta\mu_1,$$

for $a, b, c \in A$, which is exactly the statement that μ_1 is a cocycle in the Hochschild complex of A with coefficients in A.

Definition 4.5. An *infinitesimal deformation* is a deformation as given in definition 4.1 over the Artinian k-algebra of *dual numbers* $k[t]/(t^2)$.

Remark 4.5. In analogy to lemma 4.1, an infinitesimal deformation of A is given by a k-linear map $\mu_1 : A \otimes A \to A$ satisfying

$$a\mu_1(b, c) - \mu_1(ab, c) + \mu_1(a, bc) - \mu_1(a, b)c = 0,$$

for all $a, b, c \in A$. By remark 4.4, an infinitesimal deformation of A is simply a cocycle in the Hochschild complex of A with coefficients in A.

Lemma 4.2. *There is a one-to-one correspondence between the space of equivalence classes of infinitesimal deformations of A and the second Hochschild cohomology group $H^2_{Hoch}(A, A)$.*

Proof. (Sketch of proof) Two infinitesimal deformations of A are given by multiplications μ and μ' respectively. We have

$$\mu(a, b) = ab + t\mu_1(a, b) \quad \text{and} \quad \mu'(a, b) = ab + t\mu'_1(a, b),$$

for all $a, b \in A$. One can show that each equivalence of deformations $\phi : (B, \mu) \to (B, \mu')$ is determined by its restriction to A. Thus, an equivalence is given by a k-linear map $\phi_1 : A \to A$

$$\phi(a) = a + t\phi_1(a),$$

for $a \in A$ and moreover, ϕ is invertible. Since ϕ is a morphism of algebras we have

$$\phi(\mu(a, b)) = \mu'(\phi(a), \phi(b)).$$

Denote by δ is the coboundary of the Hochschild complex with coefficients in A. From the above we finally get after a short computation

$$\mu_1(a,b) = \delta\phi_1(a,b) + \mu_1'(a,b).$$

□

Formal Deformations

We give some more classical results for deformations of associative algebras. Proofs will be omitted and we refer to [MDZ07] for more details.

Lemma 4.3. *Let A be an associative k-algebra such that $H^2_{Hoch}(A,A) = 0$. Then all formal deformations of A are equivalent.*

Definition 4.6. An *n-deformation* is a deformation as given in definition 4.1 over the Artinian algebra $k[t]/(t^{n+1})$.

Lemma 4.4. *An n-deformation of A is given by a family of k-linear maps*

$$\{\mu_i : A \otimes A \to A \mid 1 \leq i \leq n\},$$

subjected to the associativity condition $(\star)$ for $1 \leq m \leq n$.

Definition 4.7. An $(n+1)$-deformation of A given by $\{\mu_1, \dots, \mu_n, \mu_{n+1}\}$ is called an *extension* of the n-deformation given by $\{\mu_1, \dots, \mu_n\}$.

Lemma 4.5. *For any n-deformation, the cochain*

$$D_n := \sum_{i+j=n+1,\ i,j>0} \mu_i(a,(\mu_j(b,c)) - \mu_i(\mu_j(a,b),c) \in C^3_{Hoch}(A,A)$$

is a cocycle, that is $\delta D_n = 0$. Moreover, D_n is a coboundary, i.e. $[D_n] = 0$ in $H^3_{Hoch}(A,A)$, if and only if the n-deformation $\{\mu_1, \dots, \mu_n\}$ extends into some $(n+1)$-deformation.

Remark 4.6. By the previous lemma we see that the obstruction to extend a deformation of A is an element of $H^3_{\mathrm{Hoch}}(A,A)$, and we can extend a deformation if this element is zero.

The general picture in classical deformation theory will be analogous. Formal deformations are represented by some power series, with infinitesimal deformations being represented by a cocycle for some appropriate cohomology theory, and the obstruction to extending a deformation is that some certain cocycle is a coboundary.

4.2 Differential Graded Lie Algebras and the Maurer-Cartan Equation

We review the notion of dg Lie algebras and introduce in this context the *Maurer-Cartan equation.* The goal is to give a first idea of the key role played by the Maurer-Cartan elements of certain dg Lie algebras in the theory of formal deformations. In doing so, we encounter a first example of a *deformation functor* picking out the Maurer-Cartan elements of a given dg Lie algebra. For an expository introduction to the use of dg Lie algebras in deformation theory we suggest the lectures of M. Manetti [Man09], which are the main reference for the following.

4.2.1 Differential Graded Lie Algebras

Definition 4.8. A *graded Lie algebra* is a graded vector space $\mathfrak{g}$ endowed with a degree 0 bracket

$$[-,-]:\mathfrak{g}\otimes\mathfrak{g}\to\mathfrak{g},$$

satisfying

– *(Graded antisymmetry)*

$$[x,y]=-(-1)^{|x||y|}[y,x];$$

– *(Graded Jacobi identity)*

$$[x,[y,z]]+(-1)^{|x|(|y|+|z|)}[y,[z,x]]+(-1)^{|z|(|x|+|y|)}[z,[x,y]]=0,$$

for all homogeneous elements $x,y,z\in\mathfrak{g}$.

Definition 4.9. Let $\mathfrak{g}$ and $\mathfrak{h}$ be graded Lie algebras. A *morphism of graded Lie algebras* $f:\mathfrak{g}\to\mathfrak{h}$ is a morphism of the underlying graded vector space that is compatible with the bracket

$$f([x,y])=[f(x),f(y)],$$

for all $x,y\in\mathfrak{g}$.

Definition 4.10. A *differential graded Lie algebra* (dg Lie algebra) is a graded Lie algebra $\mathfrak{g}$ equipped with a differential d, satisfying

– (*Graded Leibniz rule*)

$$d[a,b] = [da,b] + (-1)^{|a|}[a,db],$$

for all homogeneous elements $a, b \in \mathfrak{g}$.

Definition 4.11. Let $(\mathfrak{g}, d_{\mathfrak{g}})$ and $(\mathfrak{h}, d_{\mathfrak{h}})$ be two dg Lie algebras. A *morphism of dg Lie algebras* $f : (\mathfrak{g}, d_{\mathfrak{g}}) \to (\mathfrak{h}, d_{\mathfrak{h}})$ is a morphism of graded Lie algebras that is compatible with the differentials

$$d_{\mathfrak{h}} \circ f = f \circ d_{\mathfrak{g}}.$$

Remark 4.7. Dg Lie algebras and the morphisms between them define the category **dgLieAlg**.

Example 4.1. Let (A, μ) be a commutative graded algebra and $(\mathfrak{g}, [-,-], d_{\mathfrak{g}})$ a dg Lie algebra. The tensor product $\mathfrak{g} \otimes A$ is a dg Lie algbera with bracket and differential defined by

$$[x \otimes a, y \otimes b] := (-1)^{|a||y|}[x,y] \otimes \mu(a,b), \quad d(x \otimes a) := dx \otimes a,$$

for all $a, b \in A$ and $x, y \in \mathfrak{g}$.

Dg Lie Algebra Structure on Hochschild Cochains

We need the following definitions, taken from [MDZ07].

Definition 4.12. Let V be a vector space and let $f \in \mathrm{Hom}_{\mathbf{Vect}_k}(V^{\otimes n}, V)$ and $g \in \mathrm{Hom}_{\mathbf{Vect}_k}(V^{\otimes m}, V)$ be two k-linear maps. Define $f \circ g \in \mathrm{Hom}_{\mathbf{Vect}_k}(V^{\otimes m+n-1}, V)$ by

$$\begin{aligned} &f \circ g(v_1 \otimes \cdots \otimes v_{n+m-1}) \\ &:= \sum_{i=0}^{n} (-1)^{(m-1)(i-1)} f(v_1 \otimes \cdots \otimes v_{i-1} \otimes g(v_i \otimes \cdots \otimes v_{i+m-1}) \otimes \cdots \otimes v_{m+n-1}), \end{aligned}$$

and define the bracket $[-,-]$ by

$$[f,g] := f \circ g - (-1)^{(n-1)(m-1)} g \circ f.$$

Remark 4.8. The bracket $[-,-]$ is called the *Gerstenhaber bracket.*

Consider an associative k-algebra A with underlying vector space V. By definition 4.4 we have

$$C^n_{\mathrm{Hoch}}(A, A) = \mathrm{Hom}_{\mathbf{Vect}_k}(V^{\otimes n}, V).$$

One can show that the Hochschild complex of A with coefficients in A together with the Gerstenhaber bracket carries the structure of a dg Lie algebra. More precisely we have the following.

Proposition 4.1. *The shifted Hochschild cochain complex $C^\bullet_{\mathrm{Hoch}}(A, A)[-1]$ of an associative k-algebra A has the structure of a dg Lie algebra with respect to the Gerstenhaber bracket $[-,-]$.*

Remark 4.9. The shift in the Hochschild complex is needed for the Gerstenhaber bracket to be a degree 0 bracket.

4.2.2 The Maurer-Cartan Equation

Definition 4.13. Let $(\mathfrak{g}, [-,-], d)$ be a dg Lie algebra. A degree 1 element $x \in \mathfrak{g}^1$ is *Maurer-Cartan* if it satisfies the *Maurer-Cartan equation*

$$dx + \frac{1}{2}[x, x] = 0.$$

Let $(\mathfrak{g}, [-,-], d)$ be a dg Lie algebra and let $k[t]/t^n$ be the Artinian algebra with ideal (t) generated by t. According to example 4.1, the tensor product

$$L := \mathfrak{g} \otimes (t)$$

has a natural structure of a dg Lie algebra. We write elements of degree m as expressions of the form

$$x_1 t + x_2 t^2 + \cdots + x_{n-1} t^{n-1}, \ x_i \in \mathfrak{g}^m.$$

Then, a degree 1 element is Maurer-Cartan if its components $x_i \in \mathfrak{g}^1$ satisfy for each $1 \leq l \leq n-1$

$$dx_l + \frac{1}{2} \sum_{i+j=l} [x_i, x_j] = 0. \qquad (\star\star)$$

Remark 4.10. We denote the set of Maurer-Cartan elements in $L^1 = \mathfrak{g}^1 \otimes (t)$ by $\mathrm{MC}_{\mathfrak{g}}(k[t]/t^n)$. More generally, for every dg Lie algebra $\mathfrak{g}$ we have a functor

$$\mathrm{MC}_{\mathfrak{g}} : \mathbf{Art}_k \to \mathbf{Set},$$

called the *Maurer-Cartan functor* associated to $\mathfrak{g}$, defined by

$$\mathrm{MC}_{\mathfrak{g}}(A) := \{x \in \mathfrak{g}^1 \otimes \mathfrak{m}_A \mid x \text{ is Maurer-Cartan}\},$$

where $\mathfrak{m}_A$ is the maximal ideal of the Artinian algebra A. As it turns out, this is a *deformation functor.*

Example 4.2. Let $\mathfrak{g} = C^\bullet_{\mathrm{Hoch}}(A, A)[-1]$, together with the Gerstenhaber bracket $[-,-]$, be the dg Lie algebra structure on the shifted Hochschild cochains. Define the dg Lie algebra $L := \mathfrak{g} \otimes (t)$ and write degree 1 elements in the form

$$\mu_1 t + \mu_2 t^2 + \cdots + \mu_{n-1} t^{n-1}, \ \mu_i \in C^2_{\mathrm{Hoch}}(A, A).$$

One can check directly that the condition $(\star\star)$ for elements in L^1 to be Maurer-Cartan is equivalent to the associativity condition $(\star)$ for elements in $C^2_{\mathrm{Hoch}}(A, A)$. Hence, in this example the set $\mathrm{MC}_{\mathfrak{g}}(k[t]/t^n)$ describes formal deformations of $\mu_0 \in C^2_{\mathrm{Hoch}}(A, A)$.

4.3 Formal Description of Classical Deformation Theory

In algebraic geometry, deformation problems are conveniently formulated via functors that behave like the formal neighborhood of a point. The goal is to motivate this viewpoint and to formalize classical deformation problems via *deformation functors*, following [CG16] and [Man09].

4.3.1 Motivation and First Examples

Let $\mathcal{S}$ be some category of spaces and say we want to describe a particular space $X \in \mathcal{S}$. Recall the Yoneda embedding

$$\begin{aligned} H_\bullet : \mathcal{S} &\to [\mathcal{S}^{\mathrm{op}}, \mathbf{Set}] \\ X &\mapsto H_X : Y \mapsto \mathrm{Hom}_{\mathcal{S}}(Y, X). \end{aligned}$$

In words, instead of studying the space X directly we study how other spaces map into it. Thus, we have to consider the totality of maps into X, which is exactly provided by the functor H_X.

Formal deformations describe the local structure of X around some point $p \in X$. Hence, we need to work in a category that captures the idea of small neighborhoods of a point and how they map into X around p. To that end, let $\mathcal{S} = \mathbf{Sch}_k$ be the category of schemes over k.

Remark 4.11. Let $X \in \mathbf{Sch}_k$ and consider the functor H_X. In fact, it is enough to restrict H_X to the category of affine schemes $\mathbf{Aff}_k$. Since $\mathbf{Aff}_k$ is the opposite category to $\mathbf{cAlg}_k$ we have that every scheme X provides a functor

$$H_X : \mathbf{cAlg}_k \to \mathbf{Set}.$$

A *point* in the category $\mathbf{Sch}_k$ is provided by the affine scheme $\mathrm{Spec}(k)$ which is a one-point space endowed with k as its algebra of functions. The notion of a *point in the scheme* X is then equivalent to a map

$$P : \mathrm{Spec}(k) \to X.$$

In this setting, *small neighborhoods* of a point are described via *local Artinian algebras.*

Definition 4.14. A *local Artinian algebra* over k is a commutative k-algebra A with unique maximal ideal $\mathfrak{m}_A$ such that

- A is finite-dimensional as a k-vector space;
- there is some integer n such that $\mathfrak{m}_A^n = 0$.

Definition 4.15. Let A and B be local Artinian algebras. A *morphism of local Artinian algebras* $f : A \to B$ is a morphism of algebras which sends $\mathfrak{m}_A$ to $\mathfrak{m}_B$.

Remark 4.12. Local Artinian algebras over k and the morphisms between them define the category $\mathbf{Art}_k$.

The idea to choose the category $\mathbf{Art}_k$ to model small neighborhoods of a point is based on the following observations. A local Artinian algebra A has a unique prime ideal and hence its spectrum consists of a single point, that is $\mathrm{Spec}(A)$ is trivial as a topological space. However, $\mathrm{Spec}(A)$ has a non-trivial algebra of functions. Moreover, we have $f^n = 0$ for $f \in \mathfrak{m}_A$ and some integer n, thus we can say that $\mathrm{Spec}(A)$ has *infinitesimal directions.* Let us look at a specific example.

Example 4.3. Consider the group $\mathrm{SL}_2(R)$ of 2×2 matrices with entries in some commutative k-algebra R and determinant equal to 1. We can view SL_2 as the functor

$$\begin{aligned} \mathrm{SL}_2 : \mathbf{cAlg}_k &\to \mathbf{Set} \\ R &\mapsto \left\{ M = \begin{pmatrix} a & b \\ c & d \end{pmatrix} \;\middle|\; a,b,c,d \in R \text{ such that } 1 = ad - bc \right\}. \end{aligned}$$

We want to find the tangent space $T_{\mathbb{1}}\mathrm{SL}_2$, corresponding to first order deformations around $\mathbb{1}$. In our setting, first order deformations are described via the *dual numbers* $k[\epsilon]/\epsilon^2 \in \mathbf{Art}_k$. The tangent space is then described by matrices

$$\begin{pmatrix} 1 + s\epsilon & t\epsilon \\ u\epsilon & 1 + v\epsilon \end{pmatrix},$$

such that $(1+s\epsilon)(1+v\epsilon) - tu\epsilon^2 = 1 + (s+v)\epsilon = 1$. Hence, first order deformations correspond to traceless 2×2 matrices

$$T_{\mathbb{1}} SL_2 \simeq \mathfrak{sl}_2.$$

Example 4.4. Once again, let us consider the standard example of deformations of associative algebras. Let V be a finite dimensional vector space over k and denote by $\mathrm{Ass}_k(V)$ the set of associative algebra structures on V, that is the set of multiplications $\mu \in \mathrm{Hom}_{\mathbf{Vect}_k}(V \otimes V, V)$ satisfying

$$\mu \circ (\mu \otimes \mathrm{id}_V) = \mu \circ (\mathrm{id}_V \otimes \mu).$$

There is a natural action of $\mathrm{GL}_k(V)$ on $\mathrm{Ass}_k(V)$. Namely, given an element $g \in \mathrm{GL}_k(V)$, the action $\mu \mapsto \mu_g$ is defined by

$$\mu_g(a,b) \coloneqq g(\mu(g^{-1}(a), g^{-1}(b))),$$

for all $a, b \in V$. Infinitesimal deformations of associative algebras can then be studied by the functor

$$\begin{aligned}\mathfrak{A} : \mathbf{Art}_k &\to \mathbf{Set}\\ R &\mapsto \mathrm{Ass}_R(R \otimes V)/\mathrm{GL}_R(R \otimes V).\end{aligned}$$

4.3.2 Deformation Functors

In order to formalize the above observations, let $P : \mathrm{Spec}(k) \to \mathrm{Spec}(R)$ be a point in an affine scheme corresponding to the map $p : R \to k$ and let $(A, \mathfrak{m}_A)$ be a local Artinian algebra. A *formal A-deformation* of P is a map

$$F : \mathrm{Spec}(A) \to \mathrm{Spec}(R),$$

which sends the unique closed point in $\mathrm{Spec}(A)$ to P. Pictorially, we have the following commutative diagram

$$\begin{array}{ccc} & \overset{Q}{\nearrow} & \mathrm{Spec}(A) \\ \mathrm{Spec}(k) = \{\star\} & & \downarrow F \\ & \underset{P}{\searrow} & \mathrm{Spec}(R) \end{array}$$

where $Q : \mathrm{Spec}(k) \to \mathrm{Spec}(A)$ corresponds to the map $q : A \to A/\mathfrak{m}_A \simeq k$. We obtain a functor

$$\begin{aligned}h_P : \mathbf{Art}_k &\to \mathbf{Set}\\ (A, \mathfrak{m}_A) &\mapsto \left\{f : R \to A \mid p = q \circ f\right\}.\end{aligned}$$

More generally, this motivates the following definitions of functors formalizing deformation problems. Throughout the rest of this section, we adapt the terminology of [Man09].

Definition 4.16. A *functor of Artin rings* is a functor

$$F : \mathbf{Art}_k \to \mathbf{Set},$$

such that $F(k)$ is a singleton.

Remark 4.13. We can think of $F(k)$ as the object we want to deform over the Artinian algebra A and $F(A)$ is the set of equivalence classes of deformations.

Example 4.5. Let R be a local complete k-algebra. The functor

$$h_R : \mathbf{Art}_k \to \mathbf{Set}, \quad h_R(A) := \mathrm{Hom}_{\mathbf{cAlg}_k}(R, A),$$

is a functor of Artin rings.

Definition 4.17. A functor $F : \mathbf{Art}_k \to \mathbf{Set}$ is called *prorepresentable* if it is isomorphic to the functor h_R, as defined in example 4.5, for some local complete k-algebra R.

Remark 4.14. The formal power series $k[[x_1, \dots, x_n]]$ is a local complete k-algebra, thus motivating the terminology of *formal* deformation theory.

A *classical formal moduli problem* is a functor of Artin rings satisfying additional conditions. Here, we state the definition as given in [Man09].

Definition 4.18. A *deformation functor* is a functor of Artin rings

$$F : \mathbf{Art}_k \to \mathbf{Set},$$

such that for every pullback

$$\begin{array}{ccc} B \times_A C & \longrightarrow & C \\ \downarrow & & \downarrow \\ B & \xrightarrow{\sigma} & A \end{array}$$

in $\mathbf{Art}_k$ the induced map

$$F(B \times_A C) \to F(B) \times_{F(A)} F(C)$$

- is surjective whenever σ is surjective;
- is bijective whenever $A = k$.

Example 4.6. The functor h_R defined in example 4.5 is a deformation functor since the natural map

$$h_R(B \times_A C) \to h_R(B) \times_{h_R(A)} h_R(C)$$

is bijective.

Example 4.7. The Maurer-Cartan functor $\mathrm{MC}_{\mathfrak{g}}$ of a dg Lie algebra $\mathfrak{g}$ introduced in remark 4.10 is a deformation functor.

4.4 Derived Deformation Theory

In the setting of *derived* algebraic geometry deformation theory is concerned with the interplay between homological algebra and aspects of classical deformation theory. That is, we pass from commutative algebras describing deformations to commutative dg algebras. Also, we replace the target category **Set** by the category of simplicial sets **sSet**, or more generally by a category that allows to perform homotopy theory.

In this section we first introduce the notion of a formal deformation functor in the derived setting, following [CG16]. Then, we give the definition of an L_∞-algebra and state the main theorem that guarantees that *every* formal moduli problem is represented by a certain L_∞-algebra.

4.4.1 Derived Formal Moduli Problems

Definition 4.19. An *Artinian dg algebra* over k is a commutative dg algebra R over k, concentrated in non-positive degrees, such that

- each component R^n is finite dimensional and $R^n = 0$ for $n << 0$;
- R has a unique maximal differential graded nilpotent ideal $\mathfrak{m}_R$ such that $R/\mathfrak{m}_R \simeq k$.

Definition 4.20. Let R and S be Artinian dg algebras. A *morphism of Artinian dg algebras* $f : R \to S$ is a morphism of dg algebras which sends $\mathfrak{m}_R$ to $\mathfrak{m}_S$.

Remark 4.15. Artinian dg algebras over k and the morphisms between them define the category $\mathbf{dgArt}_k$.

Here we introduce the notion of a *derived formal moduli problem* as given in [CG16].

Definition 4.21. A *derived formal moduli problem* over a field k is a functor

$$\mathfrak{F} : \mathbf{dgArt}_k \to \mathbf{sSet},$$

satisfying the following properties

- $\mathfrak{F}(k)$ is contractible;
- $\mathfrak{F}$ takes surjective maps of Artinian dg algebras to fibrations of simplicial sets;
- for $A, B, C \in \mathbf{dgArt}_k$ and $B \to A$ and $C \to A$ surjective maps, we require that the natural map

$$\mathfrak{F}(B \times_A C) \to \mathfrak{F}(B) \times_{\mathfrak{F}(A)} \mathfrak{F}(C)$$

  is a weak homotopy equivalence.

4.4.2 The Role of L_∞-Algebras in Derived Deformation Theory

The notion of an L_∞-algebra generalizes the notion of a dg Lie algebra in that the graded Jacobi identities do not hold 'on the nose' but only up to homotopy. In the following, we formalize this idea and introduce the Maurer-Cartan equation for an L_∞-algebra. We explain how the Maurer-Cartan elements of an L_∞-algebra provide a formal moduli functor and state the main theorem that ensures that in fact *every* formal moduli problem arises in that way.

L_∞-Algebras and the Maurer-Cartan Equation

Definition 4.22. An L_∞-algebra is a graded vector space $\mathfrak{g}$ endowed with a collection of linear maps of degree $(2-n)$

$$l_n : \mathfrak{g}^{\otimes n} \to \mathfrak{g},$$

called *n-ary brackets*, for $n \in \mathbb{N}$, satisfying the following

– *(Graded antisymmetry)*

$$l_n(x_1 \otimes \cdots \otimes x_i \otimes x_{i+1} \otimes \cdots \otimes x_n) = -(-1)^{|x_i||x_{i+1}|} l_n(x_1 \otimes \cdots \otimes x_{i+1} \otimes x_i \otimes \cdots \otimes x_n),$$

for all $1 \le i \le n-1$;

– *(Generalized Jacobi identity)*

$$\sum_{\substack{i,j \in \mathbb{N} \\ i+j=n+1}} (-1)^{i(j-1)} \sum_{\sigma \in \mathrm{Sh}(i,n-i)} (-1)^{|\sigma|} \epsilon(\sigma; x_1, \ldots, x_n) \times l_j(l_i(x_{\sigma(1)} \otimes \cdots \otimes x_{\sigma(i)}) \otimes x_{\sigma(i+1)} \otimes \cdots \otimes x_{\sigma(n)}) = 0,$$

for all $n \ge 1$,

for all homogeneous elements $x_1, \ldots, x_n \in \mathfrak{g}$.

Remark 4.16. $\mathrm{Sh}(n, m-n)$ is the set of *shuffles of type* $(n, m-n)$, that is, all permutations σ of m elements such that

$$\sigma(1) < \sigma(2) < \cdots < \sigma(n) \quad \text{and} \quad \sigma(n+1) < \sigma(n+2) < \cdots < \sigma(m).$$

Remark 4.17. The sign of the permutation is included by the factor $(-1)^{|\sigma|}$. The sign $\epsilon(\sigma; x_1, \ldots, x_k)$ is determined as follows. If the permutation σ interchanges i and $i+1$, then

$$\epsilon(\sigma; x_1, \ldots, x_n) = (-1)^{|x_i||x_{i+1}|}.$$

In addition, if τ is another permutation, we have

$$\epsilon(\tau\sigma; x_1, \ldots, x_n) = \epsilon(\tau; x_{\sigma(1)}, \ldots, x_{\sigma(n)}) \epsilon(\sigma; x_1, \ldots, x_n).$$

Example 4.8. Let us denote l_1 and l_2 by d and $[-,-]$ respectively. The generalized Jacobi identity for $n = 1$ reads

$$d \circ d = 0,$$

hence d is a differential. For $n = 2$ we have

$$\begin{aligned} d[x_1, x_2] &= [dx_1, x_2] - (-1)^{|x_1||x_2|}[dx_2, x_1] \\ &= [dx_1, x_2] + (-1)^{|x_1|}[x_1, dx_2], \end{aligned}$$

that is d is a derivation of degree 1 for the bracket $[-,-]$. Notice that this bracket is graded antisymmetric but it does not necessarily satisfy the graded

Jacobi identity. However, the deviation from the graded Jacobi identity is controlled by the linear map $l_3 : \mathfrak{g}^{\otimes 3} \to \mathfrak{g}$. More precisely, for $n = 3$ the generalized Jacobi identity yields

$$
\begin{aligned}
&[x_1,[x_2,x_3]] + (-1)^{|x_1|(|x_2|+|x_3|)}[x_2,[x_3,x_1]] + (-1)^{|x_3|(|x_1|+|x_2|)}[x_3,[x_1,x_2]] \\
&= dl_3(x_1 \otimes x_2 \otimes x_3) + l_3(dx_1,x_2,x_3) + (-1)^{|x_1|}l_3(x_1,dx_2,x_3) \\
&+ (-1)^{|x_1|+|x_2|}l_3(x_1,x_2,dx_3).
\end{aligned}
$$

Thus, the graded Jacobi identity for the bracket holds up to a homotopy given by l_3. Analogously, higher generalized Jacobi identities control higher homotopy properties of higher brackets l_n on $\mathfrak{g}^{\otimes n}$.

In analogy to dg Lie algebras we can characterize the Maurer-Cartan elements of an L_∞-algebra.

Definition 4.23. Let $\mathfrak{g}$ be an L_∞-algebra. A degree 1 element $x \in \mathfrak{g}$ is *Maurer-Cartan* is it satisfies the *Maurer-Cartan equation*

$$
\sum_{n=1}^{\infty} \frac{1}{n!} l_n(x^{\otimes n}) = 0.
$$

L_∞-Algebras and Derived Deformation Theory

Definition 4.24. Given an L_∞-algebra $\mathfrak{g}$, its *Maurer-Cartan functor*

$$
\mathrm{MC}^{\mathfrak{g}} : \mathbf{dgArt}_k \to \mathbf{sSet}
$$

sends $(A, \mathfrak{m}_A)$ to the simplicial set whose n-simplices are solutions to the Maurer-Cartan equation in $\mathfrak{g} \otimes \mathfrak{m}_A \otimes \Omega^\bullet(\Delta^n)$, namely

$$
\mathrm{MC}^{\mathfrak{g}}_n(A) \coloneqq \left\{\alpha \in \mathfrak{g} \otimes \mathfrak{m}_A \otimes \Omega^\bullet(\Delta^n) \mid |\alpha| = 1, d_{\mathrm{tot}} + \frac{1}{2}[\alpha,\alpha] = 0\right\},
$$

where d_{tot} denotes the differential on $\alpha \in \mathfrak{g} \otimes \mathfrak{m}_A \otimes \Omega^\bullet(\Delta^n)$. Here, $\Omega^\bullet(\Delta^n)$ is the complex of differential forms on the standard simplicial n-simplices, as introduced in appendix B. When working over $k = \mathbb{R}$, $\Omega^\bullet(\Delta^n)$ is the usual de Rham complex.

The Maurer-Cartan functor satisfies the conditions of definition 4.21.

Theorem 4.1. *([Get09]). The Maurer-Cartan functor* $\mathrm{MC}^{\mathfrak{g}}$ *is a derived formal moduli problem.*

The following theorem, due to J. Lurie, assures that in fact *all* formal moduli problems are represented, up to a notion of weak equivalence, by the Maurer-Cartan functor of an L_∞-algebra.

Theorem 4.2. *([Lur11]) For a field of characteristic 0, there is an equivalence between the $(\infty, 1)$-categories of formal pointed moduli problems and differential graded Lie algebras over k.*

Remark 4.18. Broadly speaking, an $(\infty, 1)$-category describes a class of objects, whose morphisms have a notion of 'higher morphisms', called 2-morphisms, between them, as well as 3-morphisms between these 2-morphisms and so on. For instance, for chain complexes we can think of higher morphisms as chain homotopies, or similarly for dg algebras as homotopies between dg algebra morphisms. Moreover, in an $(\infty, 1)$-category, the n-morphisms are weakly invertible for all $n > 1$, that is invertible up to higher morphisms.

By theorem 4.2 we can trade formal moduli problems with L_∞-algebras or equivalently with dg Lie algebras. Hence, geometric concepts, such as functions on moduli problems, get an algebraic description. For instance, let $B\mathfrak{g}$ be the derived formal moduli problem associated to a dg Lie algebra $\mathfrak{g}$. Functions on $B\mathfrak{g}$ are described by the Chevalley-Eilenberg cochains $CE^\bullet(\mathfrak{g})$. Or similarly, the tangent complex of $B\mathfrak{g}$ corresponds to $\mathfrak{g}[1]$. See [CG16] for details.

5 Factorization Algebras

A factorization algebra is a prefactorization algebra satisfying a *descent property* that expresses how the factorization algebra on a large open set is determined, in a precise way, by its behavior on smaller open sets. Since there is a close relation between prefactorization algebras and precosheaves, we can think of this local-to-global property as the analog of the gluing axiom for sheaves.

We begin by recalling the notion of sheaves and cosheaves. We then define the notion of a prefactorization algebra and describe what a *G-equivariant* prefactorization algebra is for topological spaces equipped with an action of a group G. Finally, we formalize the locality axiom for factorization algebras. Since we are mostly interested in factorization algebras taking values in cochain complexes, we introduce the appropriate notion of *homotopy* factorization algebras in this setting. The presented discussion is based on [CG16] and [Sto14].

5.1 Sheaves and Cosheaves

Roughly speaking, sheaves allow to locally attach data to any open subset of a topological space and glue together this local data to obtain something global. More precisely, let M be a topological space and denote with $\mathscr{A}(U)$ the data attached to some open subset $U \subset M$. Given this data is compatible with restriction to smaller subsets, one obtains the notion of a *presheaf*. In many situations it is desirable to reconstruct global information by gluing together pieces of local data. This local-to-global property is made precise by introducing the notion of a *sheaf*.

We start with a basic example, the sheaf of continuous functions on a topological space, that captures the main ideas and properties to motivate the formal definitions. We also introduce the dual notions of a *precosheaf* and a *cosheaf*, respectively.

C. Keller, *Chern-Simons Theory and Equivariant Factorization Algebras*, BestMasters, https://doi.org/10.1007/978-3-658-25338-7_5

5.1.1 Motivating Example: The Sheaf of Continuous Functions

Let M be a topological space. To every open subset $U \subset M$ we assign the vector space $C^0(U)$ of real-valued continuous functions on U. The inclusion of subsets $U \subset V$ induces a linear *restriction map*

$$r_V^U : C^0(V) \to C^0(U).$$

Let $\{U_i\}_{i \in I}$ be an open over of U. For any $i, j \in I$ we have a commutative diagram

$$\begin{array}{ccc} U & \hookleftarrow & U_i \\ \uparrow & & \uparrow \\ U_j & \hookleftarrow & U_i \cap U_j \end{array}$$

where all the arrows are given by inclusion. The induced restriction maps thus constitute a commutative diagram

$$\begin{array}{ccc} C^0(U) & \xrightarrow{r_U^{U_i}} & C^0(U_i) \\ {\scriptstyle r_U^{U_j}}\downarrow & & \downarrow{\scriptstyle r_{U_i}^{U_i,U_j}} \\ C^0(U_j) & \xrightarrow[r_{U_j}^{U_i,U_j}]{} & C^0(U_i \cap U_j) \end{array}$$

Taking products of the restriction maps, we obtain the following maps

$$C^0(U) \xrightarrow{h} \prod_{k \in I} C^0(U_k) \underset{g}{\overset{f}{\rightrightarrows}} \prod_{i,j \in I} C^0(U_i \cap U_j).$$

Commutativity of the above implies $f \circ h = g \circ h$. One can actually show that $C^0(U)$ is isomorphic to the equalizer of f and g, that is diagrammatically

$$C^0(U) \xrightarrow{\simeq} \lim\Big(\prod_{k \in I} C^0(U_k) \underset{g}{\overset{f}{\rightrightarrows}} \prod_{i,j \in I} C^0(U_i \cap U_j)\Big).$$

We say that C^0 is a *sheaf* on M taking values in the category $\mathbf{Vect}_{\mathbb{R}}$ of real vector spaces.

5.1.2 Presheaves and Sheaves

We now extract the formal properties of the previous example to define the notion of a presheaf and a sheaf, respectively. To that end, we replace $\mathbf{Vect}_{\mathbb{R}}$ by any category $\mathcal{C}$ with products to define *sheaves with values in a category* $\mathcal{C}$.

Recall that we can regard the family $\mathbf{Opens}(M)$ of open sets of a topological space M, partially ordered by inclusion, as a category, i.e. the objects are open sets in M and there is a morphism from U to V exactly if $U \subset V$, $U, V \subset M$.

Definition 5.1. Let $\mathcal{C}$ be a category with products. A *presheaf* $\mathscr{A}$ on a topological space M with values in $\mathcal{C}$ is a functor

$$\mathscr{A} : \mathbf{Opens}(M)^{\mathrm{op}} \to \mathcal{C}.$$

Remark 5.1. A presheaf assigns an object $\mathscr{A}(U)$ in $\mathcal{C}$ to every open subset $U \subset M$ and, given an open cover $\{U_i\}_{i \in I}$ of U, for every inclusion $U_i \subset U$ there is an induced *restriction map*

$$r_U^{U_i} : \mathscr{A}(U) \to \mathscr{A}(U_i).$$

Moreover, there is a canonical map

$$\mathscr{A}(U) \to \prod_{i \in I} \mathscr{A}(U_i),$$

given by taking the product of the restriction maps.

A *sheaf* is a presheaf satisfying some *local-to-global property*, that is one can reconstruct $\mathscr{A}(U)$ in terms of $\mathscr{A}(U_i)$ and $\mathscr{A}(U_i \cap U_j)$ associated to a collection of smaller subsets $\{U_i\}_{i \in I}$ of U.

Definition 5.2. A *sheaf* on a topological space M with values in a category $\mathcal{C}$ is a presheaf $\mathscr{A}$ such that for every open subset $U \subset M$ and every open cover $\{U_i\}_{i \in I}$ of U, we have

$$\mathscr{A}(U) \xrightarrow{\ \cong\ } \lim\Big(\textstyle\prod_{k \in I} \mathscr{A}(U_k) \rightrightarrows \prod_{i,j \in I} \mathscr{A}(U_i \cap U_j) \Big),$$

where the map out of $\mathscr{A}(U)$ is the product of the restriction maps for the inclusion of each U_i into U and where, in the limit diagram, the arrows are the products of restriction maps for the inclusion of $U_i \cap U_j$ into U_i and into U_j respectively.

5.1.3 Precosheaves and Cosheaves

We now dualize the above definitions.

Definition 5.3. Let $\mathcal{C}$ be a category with coproducts. A *precosheaf* $\mathscr{G}$ on a topological space M with values in $\mathcal{C}$ is a functor

$$\mathscr{G} : \mathbf{Opens}(M) \to \mathcal{C}.$$

Remark 5.2. Given an open cover $\{U_i\}_{i\in I}$ of U, there is an induced *extension map*

$$i_{U_i}^U : \mathscr{G}(U_i) \to \mathscr{G}(U),$$

for every inclusion $U_i \subset U$. Moreover, there is a canonical map

$$\coprod_{i\in I} \mathscr{G}(U_i) \to \mathscr{G}(U),$$

given by taking the coproduct of the extension maps.

Definition 5.4. A *cosheaf* on a topological space M with values in a category $\mathcal{C}$ is a precosheaf $\mathscr{G}$ such that for every open subset $U \subset M$ and every open cover $\{U_i\}_{i\in I}$ of U, we have

$$\operatorname{colim}\Big(\textstyle\coprod_{i,j\in I} \mathscr{G}(U_i \cap U_j) \rightrightarrows \coprod_{k\in I} \mathscr{G}(U_k)\Big) \xrightarrow{\cong} \mathscr{G}(U),$$

where the map into $\mathscr{G}(U)$ is the coproduct of the extension maps for the inclusion of each U_i into U and where, in the colimit diagram, the arrows are the coproducts of extension maps for the inclusion of $U_i \cap U_j$ into U_i and into U_j respectively.

5.2 Prefactorization Algebras

The notion of a prefactorization algebra resembles a precosheaf, except that we take the tensor product instead of taking the categorical coproduct of the extension maps.

Definition 5.5. *([CG16], volume 1, section 3.1.1).* Let $(\mathcal{C}, \otimes)$ be a symmetric monoidal category. A *prefactorization algebra* $\mathcal{F}$ on a topological space M with values in $\mathcal{C}$ is a functor

$$\mathcal{F} : \mathbf{Opens}(M) \to \mathcal{C},$$

such that for any finite collection $U_1, \ldots, U_k$ of disjoint open subsets of an open subset $V \subset M$ we have a morphism

$$m_V^{U_1,\ldots,U_k} : \mathcal{F}(U_1) \otimes \cdots \otimes \mathcal{F}(U_k) \to \mathcal{F}(V).$$

These maps are compatible in the obvious way, so that if $U_{i,1}, \ldots, U_{i,n_i}$ are disjoint open subsets of V_i, and $V_1, \ldots, V_k$ are disjoint open subsets of $W \subset M$, the diagram

$$\begin{array}{ccc} \bigotimes_{i=1}^k \bigotimes_{j=1}^{n_i} \mathcal{F}(U_{i,j}) & \xrightarrow{\otimes_{i=1}^k m_{V_i}^{U_{i,1},\ldots,U_{i,n_i}}} & \bigotimes_{i=1}^k \mathcal{F}(V_i) \\ & {\scriptstyle m_W^{U_{1,1},\ldots,U_{k,n_k}}} \searrow \quad \swarrow {\scriptstyle m_W^{V_1,\ldots,V_k}} & \\ & \mathcal{F}(W) & \end{array}$$

is commutative.

Example 5.1. Let F be a precosheaf of vector spaces over some space X. For example, we can think of F as the precosheaf C_c^∞ of compactly supported smooth functions on a differentiable manifold X. Then, consider the functor $\mathcal{F}$ defined by

$$\mathcal{F} \coloneqq \mathrm{Sym}(F) : U \mapsto \mathrm{Sym}(F(U)).$$

This is a prefactorization algebra on X. Indeed, let $U, V \subset X$ be two disjoint open subsets. As F is a precosheaf we have the structure maps $F(U) \to F(U \sqcup V)$ and $F(V) \to F(U \sqcup V)$. There is a canonical map

$$F(U) \oplus F(V) \to F(U \sqcup V)$$

given by the coproduct of the structure maps, hence we get a map

$$\mathrm{Sym}(F(U) \oplus F(V)) \to \mathrm{Sym}(F(U \sqcup V)),$$

but $\mathrm{Sym}(F(U) \oplus F(V)) \simeq \mathrm{Sym}(F(U)) \otimes \mathrm{Sym}(F(V))$, so we get the following map

$$\mathcal{F}(U) \otimes \mathcal{F}(V) \to \mathcal{F}(U \sqcup V),$$

which is the structure map for the prefactorization algebra $\mathcal{F}$ on X.

5.2.1 Equivariant Prefactorization Algebras

Let M be a topological space that admits an action of a group G by homeomorphisms. We introduce the notion of a *G-equivariant prefactorization algebra* on M. For every $g \in G$, we have a homeomorphism $x \mapsto g.x$ induced by the action and for each $U \subset M$ we denote by $g.U$ the subset $\{g.x \mid x \in U\}$ of M.

Definition 5.6. *([CG16], volume 1, definition 3.7.1.1).* Let $\mathcal{F}$ be a prefactorization algebra on M. We call $\mathcal{F}$ *G-equivariant* if for every $g \in G$ and every open subset $U \subset M$ we have an isomorphism

$$\sigma_g : \mathcal{F}(U) \xrightarrow{\cong} \mathcal{F}(g.U),$$

satisfying the following conditions

- For all $g, h \in G$ and all open subsets $U \in M$ we have

$$\sigma_g \circ \sigma_h = \sigma_{gh} : \mathcal{F}(U) \xrightarrow{\cong} \mathcal{F}(gh.U);$$

- Every σ_g is compatible with the multiplicative structure on $\mathcal{F}$. That is, for any finite collection $U_1, \ldots, U_n$ of disjoint open subsets of an open subset $V \subset M$, we have that the diagram

$$\begin{array}{ccc} \mathcal{F}(U_1) \otimes \cdots \otimes \mathcal{F}(U_n) & \longrightarrow & \mathcal{F}(g.U_1) \otimes \cdots \otimes \mathcal{F}(g.U_n) \\ \downarrow & & \downarrow \\ \mathcal{F}(V) & \longrightarrow & \mathcal{F}(g.V) \end{array}$$

  is commutative.

Remark 5.3. Notice that in the above definition we do not require any compatibility of the group action with a possibly smooth structure on M. In particular, if G is a Lie group acting on a smooth manifold M

via diffeomorphisms, one can define the appropriate notion of a *smoothly* G-equivariant prefactorization algebra on M, as described in [CG16].

5.3 Factorization Algebras

A *factorization algebra* is a prefactorization algebra that satisfies some kind of *local-to-global axiom.* In fact, a factorization algebra is a cosheaf with respect to a special class of covers which are called *Weiss covers.*

Definition 5.7. *([CG16], volume 1, definition 6.1.2.1).* Let M be a topological space and $U \subset M$ an open set. A collection of open sets $\mathcal{U} = \{U_i\}_{i\in I}$ is a *Weiss cover* of U if for any finite collection of points $\{x_1, \dots, x_k\} \in U$, there is an open set $U_i \in \mathcal{U}$ such that $\{x_1, \dots, x_k\} \subset U_i$.

5.3.1 Strict Factorization Algebras

Definition 5.8. *([CG16], volume 1, definition 6.1.3.1).* A *strict factorization algebra* is a prefactorization algebra $\mathcal{F}$ on M satisfying the following axioms

- *(Factorization axiom)* For any finite collection of disjoint open subsets $U_1, \dots, U_k \subset M$ with union $U := U_1 \sqcup \dots \sqcup U_k$ the morphism

$$m_U^{U_1,\dots,U_k} : \mathcal{F}(U_1) \otimes \dots \otimes \mathcal{F}(U_k) \to \mathcal{F}(U),$$

  is an isomorphism.

- *(Locality axiom)* For every open subset $U \subset M$ and every Weiss cover $\{U_i\}_{i\in I}$ of U we have

$$\operatorname{colim}\Big(\textstyle\coprod_{i,j\in I} \mathcal{F}(U_i \cap U_j) \rightrightarrows \coprod_{k\in I} \mathcal{F}(U_k)\Big) \xrightarrow{\cong} \mathcal{F}(U),$$

  that is $\mathcal{F}$ satisfies the cosheaf condition for Weiss covers.

Example 5.2. Let A be a unital associative algebra. We associate to A the prefactorization algebra A^{fact} on $\mathbb{R}$ as follows. For any open subset $I \subset \mathbb{R}$ that is the union of a finite collection of disjoint open intervals $I_1, \dots, I_k$ we assign

$$A^{\text{fact}}(I) := A^{\text{fact}}(I_1) \otimes \dots \otimes A^{\text{fact}}(I_k) := \underbrace{A \otimes \dots \otimes A}_{k}.$$

For any open subset $J \subset \mathbb{R}$, containing the open intervals $I_1, \ldots, I_k$, the structure morphism is given by multiplication in A, so we have

$$m_J^{I_1,\ldots,I_k} : A^{\text{fact}}(I_1) \otimes \cdots \otimes A^{\text{fact}}(I_k) = A \otimes \cdots \otimes A \to A^{\text{fact}}(J) = A$$
$$a_1 \otimes \cdots \otimes a_k \mapsto a_1 \cdots a_k.$$

A^{fact} is actually a strict factorization algebra.

5.3.2 Homotopy Factorization Algebras

In many situations we are interested in factorization algebras taking values in a category of homotopical nature, such as the category of cochain complexes. In such a setting we need a 'looser' notion for a factorization algebra than the one of a *strict* factorization algebra. To that end, we make the following adaptions.

- We replace the requirement of being an isomorphism by the weaker requirement of being a *quasi-isomorphism* or a *weak equivalence.*
- We work with *all* finite intersections of open subsets in the cover.
- We work with *homotopy* colimits.

Due to our interests, we focus on factorization algebras taking values in cochain complexes.

Homotopy Cosheaves

Recall that a precosheaf on a topological space M with values in **dgVect** is a functor

$$\mathscr{G} : \mathbf{Opens}(M) \to \mathbf{dgVect}.$$

Let $\mathcal{U} = \{U_i\}_{i \in I}$ be an open cover of $U \subset M$. We have a double complex $\check{C}_\bullet(\mathcal{U}, \mathscr{G}^\bullet)$, that is, we have the following commutative diagram

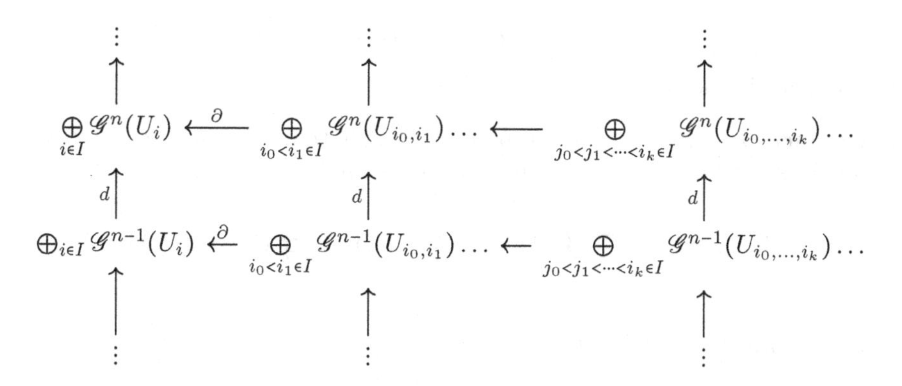

where $U_{i_0,\dots,i_k}$ denotes the k-fold intersection $U_{i_0} \cap \cdots \cap U_{i_k}$, d is the differential internal to the cochain complex assigned to each open subset by $\mathscr{G}$ and ∂ is given by the alternating sum of extension maps for the inclusion of a k-fold intersection into a $k-1$-fold intersection

$$\partial\phi_{i_0,\dots,i_k} = \bigoplus_{s=0}^{k}(-1)^s i_{U_{i_0},\dots,U_{i_k}}^{U_{i_0},\dots,\not{U}_{i_s},\dots,U_{i_k}} \phi_{i_0,\dots,i_k}.$$

The q-th column in the diagram is compactly written as $\check{C}_q(\mathcal{U},\mathscr{G}^\bullet)$. The associated *total complex* $\mathscr{C}(\mathcal{U},\mathscr{G})$ to $\check{C}_\bullet(\mathcal{U},\mathscr{G}^\bullet)$ is the complex with degree n term

$$\mathscr{C}^n(\mathcal{U},\mathscr{G}) \coloneqq \mathrm{Tot}(\check{C}(\mathcal{U},\mathscr{G}))^n \coloneqq \bigoplus_{p-q=n} \check{C}_q(\mathcal{U},\mathscr{G}^p).$$

Observe that there is a canonical map

$$\mathscr{C}(\mathcal{U},\mathscr{G}) \to \mathscr{G}(U),$$

given by taking the sum of the extension maps, i.e.

$$\bigoplus_{i\in I}\mathscr{G}(U_i) \to \mathscr{G}(U).$$

Definition 5.9. *([CG16], volume 1, A.4.3.1).* A *homotopy cosheaf* $\mathscr{G}$ on a topological space M with values in **dgVect** is a precosheaf such that for every $U \subset M$ and every open cover $\mathcal{U} = \{U_i\}_{i\in I}$ of U, we have that the natural map

$$\mathscr{C}(\mathcal{U},\mathscr{G}) \to \mathscr{G}(U).$$

is a quasi-isomorphism.

Remark 5.4. Notice that we used the result that the complex $\mathscr{C}(\mathcal{U},\mathscr{G})$ provides a representative of the homotopy colimit

$$\text{hocolim}\Big(\dots \Rrightarrow \oplus_{i<j\in I}\mathscr{G}(U_i \cap U_j) \rightrightarrows \oplus_{k\in I}\mathscr{G}(U_k)\Big)$$

Homotopy Factorization Algebra

Definition 5.10. *([CG16], volume 1, definition 6.1.4.1 and 6.1.4.2)* A *homotopy factorization algebra* on a topological space M with values in **dgVect** is a prefactorization algebra $\mathcal{F}$ satisfying the following axioms.

- *(Homotopy generalization of factorization axiom)* For every pair of disjoint open subsets $U, V \subset M$, the morphism

  $$\mathcal{F}(U) \otimes \mathcal{F}(V) \to \mathcal{F}(U \sqcup V)$$

  is a quasi-isomorphism.
- *(Homotopy generalization of locality axiom)* For any open subset $U \subset M$ and for any Weiss cover $\mathcal{U}$ of U, we have that the map

  $$\mathscr{C}(\mathcal{U},\mathcal{F}) \to \mathcal{F}(U)$$

  is a quasi-isomorphism.

Remark 5.5. Whenever we say 'factorization algebra with values in cochain complexes', we actually mean a homotopy factorization algebra.

5.3.3 The Factorization Algebra of a Classical Field Theory

We want to study observables arising from a perturbative description of field theory. Hence, as explained in chapter 4, we locally describe the derived space of solutions as a *derived formal moduli problem.* More precisely, let $\mathcal{L}$ be a sheaf of L_∞-algebras on the spacetime manifold M. Following [CG16], we define the derived formal moduli problem $B\mathcal{L}$ on $U \subset M$, parametrized by an Artinian dg algebra $(R, \mathfrak{m}_R)$, via the Maurer-Cartan functor

$$B\mathcal{L}(U)(R) = \mathrm{MC}^{\mathcal{L}(U)}(R).$$

Moreover, under the dictionary between moduli problems and L_∞-algebras, we define the classical observables $\mathrm{Obs}^{\mathrm{cl}}$ of this perturbative field theory to be the factorization algebra that assigns to each open subset U the Chevalley-Eilenberg cochain complex

$$\mathrm{Obs}^{\mathrm{cl}}(U) = CE^\bullet(\mathcal{L}(U)).$$

Results in [CG16] guarantee that this assignment indeed produces a classical factorization algebra. As pointed out in the end of chapter 4, we can view $CE^\bullet(\mathcal{L}(U))$ as the commutative dg algebra $\mathcal{O}(B\mathcal{L})$ of functions on the associated derived formal moduli problem.

Remark 5.6. Let $\mathcal{L}$ be an L_∞-algebra. The Chevalley-Eilenberg complex for cohomology $CE^\bullet(\mathcal{L})$ is the commutative dg algebra

$$\widehat{\mathrm{Sym}}(\mathcal{L}[1]^*) = \prod_{n=0}^{\infty}((\mathcal{L}[1]^*)^{\otimes n})_{S_n},$$

where $\mathcal{L}^*$ denotes the graded dual of $\mathcal{L}$.

Remark 5.7. Following [CG16], the correct notion of an L_∞-algebra defining a derived formal moduli problem is that of an *elliptic* L_∞-algebra. That is, roughly speaking, a sheaf of L_∞-algebras arising as the space of sections in a graded vector bundles over a manifold M. We will not spell out the formal definition here, but we give a basic example. The L_∞-algebra given by the de Rham complex on M

$$\Omega^\bullet_{d_{\mathrm{dR}}}(M; \mathbb{R})$$

is elliptic. As discussed in the next chapter, this is exactly the L_∞-algebra describing deformations of flat $U(1)$-bundles over M.

6 Observables in $U(1)$ Chern-Simons Theory

In the first part of this chapter, we define the Chern-Simons action for $U(1)$-bundles, following [Man98]. We describe the moduli space of gauge equivalence classes of flat $U(1)$-bundles, which turn out to be the critical points of the action functional. However, for our purposes this 'classical' moduli space of solutions is not suitable, but instead we have to take the *derived* moduli space. The local structure of the derived moduli space is described via its associated derived formal moduli problem, which gives rise to the classical factorization algebra of observables in abelian Chern-Simons theories, as introduced in [CG16]. In the last section, we construct a homotopy action of the gauge group on this classical factorization algebra, endowing the observables with an *equivariant* structure.

6.1 The Chern-Simons Functional

Let M be a closed oriented 3-manifold. Since the Lie group $U(1)$ is not simply-connected, $U(1)$-bundles over M are in general not trivializable. However, we have an inclusion homomorphism from $U(1)$ into the simply-connected Lie group $SU(2)$

$$\begin{aligned} i : U(1) &\hookrightarrow SU(2) \\ e^{2\pi i\phi} &\mapsto \begin{pmatrix} e^{2\pi i\phi} & 0 \\ 0 & e^{-2\pi i\phi} \end{pmatrix}, \end{aligned}$$

together with the corresponding inclusion of Lie algebras $i_* : \mathfrak{u}(1) \hookrightarrow \mathfrak{su}(2)$. In the following, we make use of this inclusion homomorphism and of the trivializability of $SU(2)$-bundles over M in order to define the Chern-Simons functional for connections on $U(1)$-bundles.

C. Keller, *Chern-Simons Theory and Equivariant Factorization Algebras*, BestMasters, https://doi.org/10.1007/978-3-658-25338-7_6

6.1.1 Induced Principal Bundles and Connections

Let $P \to M$ be a principal $U(1)$-bundle. We construct an $SU(2)$-bundle $P' \to M$, for which P' is the *induced principal bundle* defined by extending the structure group, i.e. it is the bundle

$$P' := P \times_{U(1),i} SU(2)$$

associated to P via the inclusion homomorphism i. Notice that there is a natural morphism of principal bundles covering the identity on M

$$i_P : P \to P', \quad p \mapsto [p, e],$$

where e is the identity element in $SU(2)$. Recall that $[p, a]$ denotes the $U(1)$-orbit through the point $(p, a) \in P \times SU(2)$. Furthermore, given a connection 1-form ω on P, there is an *induced connection* ω' on P'. Namely, the induced connection is defined by a connection 1-form ω' on $P \times SU(2)$

$$\omega'_{(p,a)} := \mathrm{ad}_{a^{-1}} \circ (i_* \mathrm{pr}_1^* \omega_p) + \mathrm{pr}_2^* \theta'_a,$$

where $\mathrm{pr}_1 : P \times SU(2) \to P$ and $\mathrm{pr}_2 : P \times SU(2) \to SU(2)$ are the natural projections and θ' is the Maurer-Cartan form on $SU(2)$. One can easily show that ω' descends to an $\mathfrak{su}(2)$-valued 1-form on the quotient of $P \times SU(2)$ by $U(1)$ and that it actually defines a connection on the induced bundle P'. The induced connection satisfies

$$i_P^* \omega' = i_* \circ \omega$$

and similarly, the curvature forms $\Omega = d\omega$ and $\Omega' = d\omega' + \frac{1}{2}[\omega' \wedge \omega']$ are related by

$$i_p^* \Omega' = i_* \circ \Omega.$$

6.1.2 The Chern-Simons Functional for $U(1)$-Connections

We now construct the Chern-Simons functional for the $U(1)$-theory. To that end, we choose a symmetric invariant bilinear form on the Lie algebra $\mathfrak{su}(2)$

$$\begin{aligned}\langle -, - \rangle : \mathfrak{su}(2) \times \mathfrak{su}(2) &\to \mathbb{R} \\ (\alpha, \beta) &\mapsto \langle \alpha, \beta \rangle := \frac{1}{8\pi^2} \mathrm{Tr}(\alpha\beta),\end{aligned}$$

where the normalization is such that the Chern-Simons functional is gauge invariant modulo 1.

Definition 6.1. *([Man98], definition 3.1).* Let $P \to M$ be a $U(1)$-bundle endowed with a connection ω. The *Chern-Simons action* $\mathcal{S}_{M,P}$ of ω is defined as

$$\mathcal{S}_{M,P}(\omega) := \int_M s'^* \alpha(\omega') \pmod 1,$$

where $\alpha(\omega')$ is the Chern-Simons 3-form of the induced connection ω' on the $SU(2)$-bundle $P' = P \times_{U(1),i} SU(2)$ and $s' : M \to P'$ is a global section.

In particular, the $U(1)$ Chern-Simons functional is invariant under gauge transformations.

Proposition 6.1. *([Man98], proposition 3.5). Let $\mathcal{A}(P)$ denote the space of connections on a principal $U(1)$-bundle $P \to M$. The Chern-Simons action functional*

$$\mathcal{S}_{M,P} : \mathcal{A}(P) \to \mathbb{R}/\mathbb{Z}$$

is invariant under the action of the group of gauge transformations $\mathcal{G}(P)$, that is,

$$\mathcal{S}_{M,P}(\omega^u) = \mathcal{S}_{M,P}(\omega),$$

where ω^u denotes the gauge transformed connection for all $u \in \mathcal{G}(P)$. Hence, $\mathcal{S}_{M,P}$ descends to a functional on the quotient space $\mathcal{A}(P)/\mathcal{G}(P)$ of gauge equivalence classes of connections on P.

Our main interest is in the critical points of the action functional. The following proposition shows that for $U(1)$-bundles the equations of motion pick out flat connections.

Proposition 6.2. *([Man98], proposition 3.6). The stationary points of the $U(1)$ Chern-Simons functional $\mathcal{S}_{M,P}$ are the flat connections, that is, $\mathcal{S}_{M,P}(\omega) = 0$ if and only if $d\omega = 0$.*

Remark 6.1. The Chern-Simons action for any toroidal Lie group

$$\mathbb{T}^n = \underbrace{U(1) \times \cdots \times U(1)}_{n \text{ times}}$$

can be similarly constructed. But notice that one has to pick an appropriate invariant pairing on the Lie algebra $\mathbb{R}^n$ in order for the action to be gauge invariant in $\mathbb{R}/\mathbb{Z}$.

6.2 The Moduli Space of Flat Bundles

Let $\mathbf{Bun}_\omega^G(M)$ be the category whose objects are principal G-bundles $P \to M$ with connections and whose morphisms $\vartheta : P \to P'$ are morphisms of principal bundles covering the identity on M. The objects of $\mathbf{Bun}_\omega^G$ form a union of affine spaces

$$\mathrm{ob}(\mathbf{Bun}_\omega^G) = \bigsqcup_{\{P\}} \mathcal{A}(P),$$

where $\mathcal{A}(P)$ denotes the affine space of connections on P and $\{P\}$ is the collection of principal bundles over M. We call a pair $(P, \omega_0) \in \mathbf{Bun}_\omega^G(M)$ a *flat G-bundle* if ω_0 is a flat connection on P.

6.2.1 The Moduli Space of G-Bundles

Denote with $\widetilde{\mathbf{Bun}}_\omega^G(M)$ the category whose objects are simply the objects of $\mathbf{Bun}_\omega^G(M)$, but whose morphisms are *connection preserving*, i.e. morphisms are principal bundle map $\vartheta : P' \to P$ satisfying

$$\omega' = \vartheta^* \omega.$$

Definition 6.2. Two objects $\omega, \omega' \in \widetilde{\mathbf{Bun}}_\omega^G(M)$ are called *gauge equivalent* if there exists a morphism between them.

Remark 6.2. If two connections ω, ω' are gauge equivalent we write $\omega \sim \omega'$. This is an equivalence relation on $\widetilde{\mathbf{Bun}}_\omega^G(M)$.

Let $\mathcal{M}(M)$ denote the set of gauge equivalence classes of bundles with connections over M and let $\{[P]\}$ be a set of representatives. Then we have that

$$\mathcal{M}(M) \simeq \bigsqcup_{\{[P]\}} \mathcal{A}(P)/\mathcal{G}(P),$$

and we call $\mathcal{M}(M)$ the *moduli space of connections* on principal G-bundles over M.

Remark 6.3. Let $\vartheta \in \mathcal{G}(P)$ be a bundle automorphisms $\vartheta : P \to P$ with associated G-equivariant map $\gamma_\vartheta : P \to G$. The action of $\mathcal{G}(P)$ on the affine space of connection $\mathcal{A}(P)$ is given by

$$\begin{aligned} \mathcal{A}(P) \times \mathcal{G}(P) &\to \mathcal{A}(P) \\ (\omega, \vartheta) &\mapsto \vartheta^* \omega = \mathrm{ad}_{\gamma_\vartheta^{-1}} \circ \omega + \gamma_\vartheta^* \theta, \end{aligned}$$

where θ is the Maurer-Cartan form on G.

6.2.2 Abelian Bundles and the Moduli Space of Flat Connections

We consider principal G-bundles with G an abelian structure group.

The Group of Gauge Transformations

First notice that in the abelian case the adjoint bundle is canonically isomorphic to the trivial G-bundle

$$\mathrm{Ad}(P) := P \times_{(G,\mathrm{Ad})} G \simeq M \times G.$$

Moreover, recall from chapter 2 that the gauge group $\mathcal{G}(P)$ of bundle automorphisms can be identified with the space of sections $\Gamma(M; \mathrm{Ad}(P))$ in the adjoint bundle. Hence, we have a canonical isomorphism

$$\mathcal{G}(P) \simeq \mathrm{Maps}(M; G),$$

for all principal bundles P over M. As a consequence, there is an action of $\mathcal{G}$ on the space of connections over M

$$\mathcal{A}(M) := \bigsqcup_{\{P\}} \mathcal{A}(P) \circlearrowleft \mathcal{G} \simeq \mathrm{Maps}(M; G).$$

Given an element $(P, \omega) \in \mathcal{A}(M)$, the $\mathcal{G}$-action is described as

$$(P, \omega) \mapsto (P, \omega)^u := (\vartheta_u(P), \vartheta_u^{-1^*} \omega)$$

where $\vartheta_u : P \to P$ is the bundle automorphism associated to the element $u \in \mathrm{Maps}(M;G)$ and the gauge transformed connection is given by

$$\vartheta_u^{-1^*}\omega = \omega + (u \circ \pi)^*\theta,$$

where θ is the Maurer-Cartan form on G.

Remark 6.4. We can describe the action

$$\begin{aligned} \mathcal{A}(M) \times \mathcal{G} &\to \mathcal{A}(M) \\ ((P,\omega),u) &\mapsto (P,\omega)^u \end{aligned}$$

via the *transformation groupoid* $\mathcal{A}(M)//\mathcal{G}$, where

- objects are elements $(P,\omega) \in \mathcal{A}(M)$;
- morphisms are pairs $((P,\omega),u) : (P,\omega) \mapsto (P,\omega)^u$;
- composition of $((P,\omega),u) : (P,\omega) \mapsto (P,\omega)^u = (P',\omega')$ and $((P',\omega'),u') : (P',\omega') \mapsto (P',\omega')^{u'}$ is given by

$$((P,\omega),u'u) : (P,\omega) \mapsto (P,\omega)^{u'u}.$$

Notice that the transformation groupoid $\mathcal{A}(M)//\mathcal{G}$ is similar to the orbit space of the $\mathcal{G}$-action, but instead of taking elements in the same $\mathcal{G}$-orbit as being equal, in the transformation groupoid they are just isomorphic.

The Moduli Space of Flat $U(1)$-Bundles

We characterize the classical moduli space of gauge equivalence classes of flat $U(1)$-connections over a base manifold M, based on results in [Man98].

Remark 6.5. The set of isomorphism classes of $U(1)$-bundles over M is isomorphic to the cohomology group $H^2(M;\mathbb{Z})$.

Remark 6.6. One can show that a $U(1)$-bundle P admits flat connections if and only if its first Chern class $c_1(P)$ is a Torsion class, that is if $c_1(P) \in \mathrm{Tors}\, H^2(M;\mathbb{Z})$.

Denote with $\mathcal{A}(P)_{\text{flat}} \subset \mathcal{A}(P)$ the space of flat connections on a principal bundle P. The space of flat connections $\mathcal{A}(M)_{\text{flat}}$ over M is then given by the union

$$\mathcal{A}(M)_{\text{flat}} = \bigsqcup_{\substack{\{P\} \\ c_1(P)\in \text{Tors } H^2(M;\mathbb{Z})}} \mathcal{A}(P)_{\text{flat}}$$

over all principal bundles whose first Chern class is a Torsion class. Notice that the curvature is invariant under abelian gauge transformations and hence, the $\mathcal{G}$-action preserves the space of flat connections. For each Chern class $c \in \text{Tors } H^2(M;\mathbb{Z})$, let us denote the corresponding space of flat connections by

$$\mathcal{A}(M)^c_{\text{flat}} = \bigsqcup_{\substack{\{P\} \\ c_1(P)=c}} \mathcal{A}(P)_{\text{flat}}.$$

Then, the moduli space of flat connections on M with first Chern class equal to c is given by $\mathcal{M}(M)^c_{\text{flat}} = \mathcal{A}(M)^c_{\text{flat}}/\sim$. Hence, we have that the *moduli space $\mathcal{M}(M)_{flat}$ of gauge equivalence classes of flat connections* on M is described as the disjoint union

$$\mathcal{M}(M)_{\text{flat}} = \bigsqcup_{c\in \text{Tors } H^2(M;\mathbb{Z})} \mathcal{M}(M)^c_{\text{flat}}.$$

Remark 6.7. Notice that for each P with $c_1(P) = c$ we have a natural isomorphism $\mathcal{A}(P)_{\text{flat}}/\mathcal{G} \simeq \mathcal{M}(M)^c_{\text{flat}}$. More precisely, let P, P' be two principal bundles with $c_1(P) = c_1(P') \in \text{Tors } H^2(M;\mathbb{Z})$. We have a canonical isomorphism

$$\mathcal{A}(P)_{\text{flat}}/\mathcal{G} \simeq \mathcal{A}(P')_{\text{flat}}/\mathcal{G}.$$

Indeed, let $\vartheta : P \to P'$ be a bundle isomorphism. This induces an isomorphism on the space of connections

$$\begin{aligned} \vartheta^* : \mathcal{A}(P') &\to \mathcal{A}(P) \\ \omega' &\mapsto \vartheta^*\omega'. \end{aligned}$$

This isomorphism descends to an isomorphism between the quotients by the gauge group. Moreover, it is independent of the choice of bundle isomorphism.

Proposition 6.3. *([Man98], proposition 2.2). There is a natural identification*

$$\mathcal{M}(M)_{flat} \simeq H^1(M;U(1)).$$

Proof. (Sketch of proof - for details of the following assertions see for example [KN96]). Let $P \to M$ be a principal G-bundle. Recall that given a connection

$\omega \in \mathcal{A}(P)$ and a loop γ in M based at $x \in M$, the *holonomy* around γ is an equivariant map

$$\mathrm{hol}_{\omega,x}(\gamma) : \pi^{-1}(\{x\}) \to \pi^{-1}(\{x\}),$$

defined by parallel transport along γ. For a fixed $p \in \pi^{-1}(\{x\})$, the holonomy can be interpreted as a Lie group element. Indeed, we can define

$$\mathrm{hol}_{\omega,x,p}(\gamma) \in G, \quad \mathrm{hol}_{\omega,x}(\gamma)p := p.\mathrm{hol}_{\omega,x,p}(\gamma).$$

A standard result in the theory of connections guarantees that a *flat* G-connection $\omega \in \mathcal{A}(P)$ determines a group homomorphism

$$\begin{aligned}\pi_1(M,x) &\to G\\ [\gamma] &\mapsto \mathrm{hol}_{\omega,x,p}(\gamma).\end{aligned}$$

The holonomy map behaves under gauge transformations as follows

$$\mathrm{hol}_{\vartheta_u^*\omega,x,p}(\gamma) = u(p)^{-1}\mathrm{hol}_{\omega,x,p}(\gamma)u(p),$$

for any gauge group element $u \in \mathrm{Maps}(P;G)^{(G,\mathrm{Ad})}$. Hence, the holonomy induces a map on the quotients

$$\mathrm{hol}_{x,p} : \mathcal{A}(P)_{\mathsf{flat}}/\mathcal{G}(P) \to \mathrm{Hom}(\pi_1(M,x),G)/G,$$

where G acts by conjugation. Moreover, one can show that the induced map $\mathrm{hol}_{x,p}$ is independent of the choices of $x \in M$ and $p \in \pi^{-1}(\{x\})$. In particular, for the abelian group $G = U(1)$ the holonomy induces a bijection

$$\mathcal{M}(M)_{\mathsf{flat}} \simeq \mathrm{Hom}(\pi_1(M),U(1)).$$

Notice that the first homology group is the abelianization of the fundamental group and the assertion follows. □

Proposition 6.4. *([Man98], proposition 2.2). Each connected component of $\mathcal{M}(M)_{flat}$ is isomorphic to the torus*

$$\mathcal{M}(M)^c_{flat} \simeq H^1(M;\mathbb{R})/H^1(M;\mathbb{Z}).$$

Proof. (Sketch of proof) Consider the short exact sequence

$$0 \to \mathbb{Z} \to \mathbb{R} \to U(1) \to 0.$$

There is an induced long exact sequence in cohomology

$$0 \to H^1(M;\mathbb{Z}) \to H^1(M;\mathbb{R}) \to H^1(M;U(1)) \xrightarrow{\delta} H^2(M;\mathbb{Z}) \xrightarrow{\iota} H^2(M;\mathbb{R}) \to \ldots,$$

with $\mathrm{Im}(\delta) = \mathrm{Ker}(\iota) = \{c | c \in \mathrm{Tors}\, H^2(M;\mathbb{Z})\}$. Hence, for each torsion class $c \in H^2(M;\mathbb{Z})$ we have

$$\mathcal{M}(M)^c_{\text{flat}} \cong \delta^{-1}(c) \cong \delta^{-1}(0) \cong H^1(M;\mathbb{R})/\iota(H^1(M;\mathbb{Z})),$$

where the map $\iota : H^1(M;\mathbb{Z}) \to H^1(M;\mathbb{R})$ is induced by inclusion. □

6.3 The Derived Formal Moduli Problem for Flat Abelian Bundles

We give the derived formal moduli problem emerging from deformations of flat connections. Our discussion is based on [CG16].

6.3.1 Deformations of Flat Bundles

Let $(P,\omega_0) \in \mathbf{Bun}^G_\omega(M)$ be a flat G-bundle. Let us denote with $\mathrm{Def}_{(P,\omega_0)}$ the moduli problem describing the local structure of the derived moduli space of flat connections around (P,ω_0). More precisely, the functor $\mathrm{Def}_{(P,\omega_0)}$ assigns to any $(R,\mathfrak{m}_R) \in \mathbf{dgArt}_k$ the simplicial set $\mathrm{Def}_{(P,\omega_0)}(R)$ of families of G-bundles, parametrized by $\mathrm{Spec}(R)$, that deform (P,ω_0). Explicitly, following [CG16], we consider the functor

$$\begin{aligned}\mathrm{Def}^n_{(P,\omega_0)}(R) :=\{& A \in \Omega^\bullet_{d_{\omega_0}}(M,\mathrm{ad}(P)) \otimes \mathfrak{m}_R \otimes \Omega^\bullet(\Delta^n) \mid \\ & |A| = 1, d_{\omega_0}A + d_R A + d_{\Delta^n} A + \frac{1}{2}[A \wedge A] = 0\},\end{aligned}$$

assigning to any Artinian dg algebra $(R,\mathfrak{m}_R)$ the Maurer-Cartan elements in $\Omega^\bullet_{d_{\omega_0}}(M;\mathrm{ad}(P)) \otimes \mathfrak{m}_R$. In other words, the formal moduli problem is represented by the dg Lie algebra

$$\mathcal{L} = \Omega^\bullet_{d_{\omega_0}}(M,\mathrm{ad}(P)),$$

where d_{ω_0} is the exterior covariant derivative induced by the flat connection ω_0.

Let us motivate the above in a non-derived setting, i.e. for $(R, \mathfrak{m}_R) \in \mathbf{Art}_k$. Recall that the difference between two connections on P is a 1-form on M with values in the adjoint bundle. Thus, a deformation of ω_0 on P, parametrized by $\mathrm{Spec}(R)$, is an element

$$A \in \Omega^1(M; \mathrm{ad}(P)) \otimes \mathfrak{m}_R.$$

The curvature of the deformed connection $\omega_0 + A$ is

$$F = d_{\omega_0} A + \frac{1}{2}[A \wedge A] \in \Omega^2(M; \mathrm{ad}(P)) \otimes \mathfrak{m}_R.$$

Hence, the formal moduli problem returns flat deformations.

6.3.2 Flat Abelian Bundles

First, recall from chapter 2 that in the abelian case the covariant exterior derivative d_{ω_0} reduces to the ordinary de Rham differential d. Moreover, we have a canonical isomorphism

$$\mathrm{ad}(P) := P \times_{(G,\mathrm{ad})} \mathfrak{g} \simeq M \times \mathfrak{g}.$$

Hence, for abelian bundles we have the identification

$$\Omega^\bullet_{d_{\omega_0}}(M; \mathrm{ad}(P)) \simeq \Omega^\bullet_d(M; \mathfrak{g}).$$

In summary, we find the following.

Fact 6.1. *The formal neighborhood of the derived space of solutions of flat abelian G-bundles over a manifold M is controlled at any of its points by the dg Lie algebra*

$$\mathcal{L} = \Omega^\bullet(M; \mathfrak{g})$$

given by the de Rham complex of $\mathfrak{g}$-valued forms on M.

Remark 6.8. Recall from chapter 5 that K. Costello and O. Gwilliams [CG16] define the classical observables $\mathrm{Obs}^{\mathrm{cl}}$ of the theory as the factorization algebra that assigns to every open subset $U \subset M$ the Chevalley-Eilenberg cochain complex

$$\mathrm{Obs}^{\mathrm{cl}}(U) = CE^{\bullet}(\mathcal{L}(U))$$

of the dg Lie algebra $\mathcal{L}$ controlling the formal deformation problem.

For abelian Chern-Simons theories we thus have that the factorization algebra of classical observables detects only the underlying Lie algebra of the theory. For instance, by fact 6.1, the observables in $\mathbb{R}$ Chern-Simons theory carry the same structure as in the $U(1)$-theory. But following [CG16], a linear observable in $\mathbb{R}$ Chern-Simons theory is given by $\int_\gamma A$ for γ a loop and A a connection 1-form on M, however, this is in general *not* an observable in the $U(1)$-theory. This suggests that the construction of classical observables has to be refined, as described in the following section.

6.4 Equivariant Factorization Algebras

We construct an action of the gauge group on the classical factorization of observables induced by gauge transformations. The resulting equivariant structure on the factorization algebra encodes more topological information about the underlying Lie group than in a purely perturbative description.

6.4.1 The Main Idea

By lemma 6.4, the 'classical' moduli space of flat connections in $U(1)$ Chern-Simons theory is identified with a disjoint union of tori

$$H^1(M;\mathbb{R})/H^1(M;\mathbb{Z}) \coloneqq \tilde{V}/\Lambda.$$

We now make the assumption that observables in this classical theory should arise as functions on the tangent space $\tilde{V}$ that are invariant under the action of the lattice Λ. More precisely, observables are expected to be functions $f : \tilde{V} \to k$, satisfying

$$f(v) = f(v + \alpha),$$

for all $v \in \tilde{V}$ and $\alpha \in \Lambda$. We make the natural choice to work with polynomial functions $P(\tilde{V}) \subset \mathcal{O}(\tilde{V})$, or equivalently, with the symmetric algebra $\mathrm{Sym}(\tilde{V}^*)$ on the dual of $\tilde{V}$.

We want to translate these ideas in a derived setting. To that end, we substitute the tangent space $\tilde{V}$ by the tangent complex

$$V := \Omega^\bullet(M; \mathbb{R})[1]$$

to the derived formal moduli problem for flat $U(1)$-bundles. Notice that we can recover the vector space $\tilde{V}$ from the zeroth cohomology group of the cochain complex V

$$H^0_{\mathrm{dR}}(V) \simeq H^1(M; \mathbb{R}).$$

Further, we choose the gauge group $\mathrm{Maps}(M; U(1))$ as a substitute for the lattice Λ. The motivation for this choice is the following. We require to recover the lattice Λ as the homotopy set $\pi_0(\mathrm{Maps}(M; U(1)))$ of maps from M to $U(1)$. Indeed, we have

$$\pi_0(\mathrm{Maps}(M; U(1))) \simeq H^1(M; \mathbb{Z}),$$

which follows from the fact that $U(1)$ is an Eilenberg-MacLane space $K(\mathbb{Z}, 1)$.

Remark 6.9. Notice that the completed commutative dg algebra

$$\widehat{\mathrm{Sym}}(V^*) = \prod_{n=0}^{\infty} ((V^*)^{\otimes n})_{S_n}$$

is the classical factorization algebra of observables described in remark 6.8. This is the factorization algebra of abelian Chern-Simons theories studied by K. Costello and O. Gwilliams [CG16]. Throughout the following, the completion will be understood.

Motivated by the classical theory, we construct a *homotopy action* of the gauge group on $\mathrm{Sym}(V^*)$, induced by gauge transformations of elements in V. We show that this gauge transformation action can actually be promoted to an action of precosheaves valued in commutative dg algebras. Hence, the gauge transformation action endows the classical observables with the structure of an *equivariant* factorization algebra.

Remark 6.10. We want to sketch briefly what we mean by a homotopy action. Let G be a topological group. We define a simplicial group $\mathcal{C}^\bullet(G)$ with n-simplices given by

$$\mathcal{C}^n(G) := \mathrm{Maps}(|\Delta^n| \times M; G).$$

Notice that the category $\mathbf{cdgAlg}_k$ has a natural simplicial enrichment. Namely, for $A, B \in \mathbf{cdgAlg}_k$ the simplicial set $\mathrm{Maps}^\bullet(A, B)$ has n-simplices

$$\mathrm{Maps}^n(A, B) := \mathrm{Hom}_{\mathbf{cdgAlg}_k}(A, B \otimes \Omega^\bullet(\Delta^n)).$$

A homotopy action of G on a commutative dg algebra A is a map of simplicial sets

$$\mathcal{C}^\bullet(G) \to \mathrm{Maps}^\bullet(A, A),$$

together with the induced maps on the homotopy groups. Thus, defining a homotopy action on the factorization algebra of observables incorporates topological information about the underlying Lie group. For instance, if G is contractible, the action is homotopically trivial.

6.4.2 Gauge Transformation Action

Lemma 6.1. *Let V be a chain complex. An element $v \in V^n$ gives rise to a linear map $V^* \to k$ of degree n via evaluation.*

Proof. Notice that we can regard k as a cochain complex concentrated in degree 0. Define

$$< v, \psi > := \begin{cases} \psi(v), & \text{if } \psi \in (V^*)^{-n} \\ 0, & \text{otherwise.} \end{cases}$$

This map is linear. □

Let (V, d) be a cochain complex and let $\Lambda \subset V$ be a subspace. There is a natural action of Λ on V given by translation

$$V \times \Lambda \to V, \quad (v, \alpha) \mapsto v + \alpha.$$

We can extend this action to $\mathrm{Sym}(V^*)$ as follows. Let α be a *closed, degree 0* element in Λ and define a map

$$\varphi_\alpha : V^* \to \mathrm{Sym}(V^*), \quad \varphi_\alpha(\psi)(v) := < v + \alpha, \psi > .$$

for all $\psi \in V^*$ and $v \in V$. This map is linear and hence, by the universal property of the symmetric algebra, φ_α induces a map $\mathrm{Sym}(\varphi_\alpha) : \mathrm{Sym}(V^*) \to$

$\mathrm{Sym}(V^*)$ for every $\alpha \in \Lambda$. This is a morphism of cochain complexes. Indeed, on generators $\psi \in V^*$ we have

$$\varphi_\alpha(\psi) = \begin{cases} \psi + \psi(\alpha), & \text{if } \psi \in (V^*)^0 \\ \psi, & \text{otherwise.} \end{cases}$$

Hence, the map is degree preserving. It also commutes with the differential.

Remark 6.11. Notice that in this context the right notion of a differential on $\varphi_\alpha(\psi) \in \mathrm{Sym}(V^*)$, for $\psi \in V^*$, is obtained by regarding it as an element of the affine dual $(\mathrm{Aff}^*(V), \bar{\delta})$, endowed with a differential defined by $\bar{\delta} f(v) := (-1)^{|f|}(f(dv) - f(0))$. Then, for $\psi \in V^*$ we have on the one hand

$$\begin{aligned} \bar{\delta}(\varphi_\alpha(\psi))(v) &= (-1)^{|\psi|}(\varphi_\alpha(\psi)(dv) - \varphi_\alpha(\psi)(0)) \\ &= (-1)^{|\psi|}(< dv + \alpha, \psi > - < \alpha, \psi >) \\ &= (-1)^{|\psi|} < dv, \psi > \end{aligned}$$

and on the other hand

$$\begin{aligned} \varphi_\alpha(\delta\psi)(v) &=< v + \alpha, \delta\psi > \\ &= (-1)^{|\psi|} < dv + d\alpha, \psi > \\ &= (-1)^{|\psi|} < dv, \psi >, \end{aligned}$$

where in the last line we used that α is closed.

Consider now $V := \Omega^\bullet(M; \mathbb{R})[1]$. We want to show that we have an action of $\mathrm{Maps}(M; U(1))$ on $\mathrm{Sym}(V^*)$. Namely, let $u \in \mathrm{Maps}(M; U(1))$ be an element of the group of gauge transformations. The pullback $u^*\theta$ of the Maurer-Cartan form θ is a closed, degree 0 element. Hence, we get a morphism of commutative dg algebras

$$\mathrm{Sym}(\varphi_{u^*\theta}) : \mathrm{Sym}(V^*) \to \mathrm{Sym}(V^*).$$

This morphism defines an action of $\mathrm{Maps}(M; U(1))$ on $\mathrm{Sym}(V^*)$. Indeed, since $U(1)$ is a matrix group we have $u^*\theta = \frac{1}{u} du$. That is, for $u, u' \in \mathrm{Maps}(M; U(1))$ we have

$$\begin{aligned}(uu')^*\theta &= \frac{1}{uu'}d(uu')\\ &= \frac{1}{u}du + \frac{1}{u'}du'\\ &= u^*\theta + u'^*\theta,\end{aligned}$$

and thus $\mathrm{Sym}(\varphi_{(uu')^*\theta}) = \mathrm{Sym}(\varphi_{u'^*\theta}) \circ \mathrm{Sym}(\varphi_{u^*\theta})$.

Action on Precosheaves

We want to show that the action constructed above is an action of precosheaves valued in commutative dg algebras.

First, for any open set $U \subset M$ let $V|_U := \Omega^\bullet(U;\mathbb{R})[1]$ and denote by Sym^{V^*} the precosheaf defined as

$$\mathrm{Sym}^{V^*}(U) := \mathrm{Sym}(V^*|_U).$$

The *extension maps* are given by $\mathrm{Sym}(\iota^!)$, where

$$\begin{aligned}\iota^! : V^*|_U &\to V^*|_W\\ \psi &\mapsto \psi \circ \iota^*\end{aligned}$$

with $\iota : U \hookrightarrow W$ the inclusion map. We have to show that each $u \in \mathrm{Maps}(M; U(1))$ induces a morphism $\mathrm{Sym}^{V^*} \to \mathrm{Sym}^{V^*}$. Namely, for any inclusion $\iota : U \hookrightarrow W$ the diagram

$$\begin{array}{ccc}\mathrm{Sym}(V^*|_U) & \xrightarrow{\mathrm{Sym}(\varphi_{u^*\theta|_U})} & \mathrm{Sym}(V^*|_U)\\ {\scriptstyle \mathrm{Sym}(\iota^!)}\downarrow & & \downarrow{\scriptstyle \mathrm{Sym}(\iota^!)}\\ \mathrm{Sym}(V^*|_W) & \xrightarrow[\mathrm{Sym}(\varphi_{u^*\theta|_W})]{} & \mathrm{Sym}(V^*|_W)\end{array}$$

has to be commutative. It is enough to check on generators. Let $\psi \in V^*|_U$, then we have for $w \in V|_W$

$$\begin{aligned}\varphi_{u^*\theta|_W}(\iota^!(\psi))(w) &= < \iota^* w + \iota^* u^*\theta|_W, \psi >\\ &= < \iota^* w + u^*\theta|_U, \psi >\\ &= \iota^!(\varphi_{u^*\theta|_U}(\psi))(w).\end{aligned}$$

Homotopies

We first recall the following useful lemma.

Lemma 6.2. *Let $f, g : V \to \mathrm{Sym}(W)$ be morphisms of chain complexes and let h be a chain homotopy between f and g. Then h induces a canonical homotopy between* $\mathrm{Sym}(f)$ *and* $\mathrm{Sym}(g)$.

Let $\alpha \in V$ be a closed degree 0 element and let $\beta \in V$ be a degree -1 element. Evaluation on β induces a homotopy between $\mathrm{Sym}(\varphi_{\alpha+d\beta})$ and $\mathrm{Sym}(\varphi_\alpha)$. Indeed, let $\psi \in V^*$ and notice that

$$\begin{aligned}(\varphi_{\alpha+d\beta} - \varphi_\alpha)(\psi) &= < d\beta, \psi > \\ &= (-1)^{|\psi|} < \beta, \delta\psi > + \delta(< \beta, \psi >) \\ &= (\delta \circ h_\beta + (-1)^{|\psi|} h_\beta \circ \delta)(\psi),\end{aligned}$$

where $h_\beta := < \beta, - >$ and we have used that $\delta \circ h_\beta$ is identically 0. By lemma 6.2, the homotopy h_β canonically induces a homotopy

$$H_\beta : \mathrm{Sym}(V^*) \to \mathrm{Sym}(V^*) \otimes k[t, dt],$$

between $\mathrm{Sym}(\varphi_{\alpha+d\beta})$ and $\mathrm{Sym}(\varphi_\alpha)$.

Consider now the case $V := \Omega^\bullet(M)[1]$. Let

$$F : M \times [0,1] \to U(1)$$

be a homotopy between $f := F(x, 0)$ and $g := F(x, 1)$. Pulling back the Maurer-Cartan form θ along F gives a 1-form $F^*\theta$ on $M \times [0,1]$. Denote by $\pi_* : \Omega^1(M \times [0,1]) \to \Omega^0(M)$ integration along the fiber. The composition $\pi_* \circ F^*$ is a homotopy operator, that is we have

$$g^* - f^* = d \circ (\pi_* \circ F^*) + (\pi_* \circ F^*) \circ d : \Omega^1(U(1); \mathbb{R}) \to \Omega^1(M; \mathbb{R}).$$

Let us denote $\beta := \pi_*(F^*\theta)$. Since θ is closed, we have

$$g^*\theta - f^*\theta = d\beta.$$

Following the results above, we thus obtain for every homotopy between $g, f \in \mathrm{Maps}(M; U(1))$ a homotopy of commutative dg algebras between $\mathrm{Sym}(\varphi_{g^*\theta})$ and $\mathrm{Sym}(\varphi_{f^*\theta})$.

6.5 Summary and Outlook

We explicitly constructed a homotopy action of the gauge group on the classical factorization algebra in $U(1)$ Chern-Simons theory, endowing the observables with an equivariant structure. Similarly, one can build a gauge transformation action on the factorization algebra in any Chern-Simons theory with an abelian group structure. Indeed, notice that every connected abelian Lie group G is isomorphic to a product

$$G \simeq \mathbb{R}^{n-k} \times \mathbb{T}^k,$$

where n is the dimension of G. By fact 6.1, for any Lie group G of the above form, the classical factorization algebra in Chern-Simons theory is controlled by the dg Lie algebra

$$\Omega^\bullet(M; \mathbb{R}^n).$$

This reflects the fact that on a purely perturbative level, the structure present on the classical observables is only sensitive to the underlying Lie algebra. However, by equipping the factorization algebra with a homotopy action of the gauge group, we can encode more topological features of the underlying Lie group in the structure of the observables and thus detect non-perturbative phenomena. For instance, take the standard example of $\mathbb{R}^n$ vs. $\mathbb{T}^n$ Chern-Simons theory. For the simply-connected Lie group $\mathbb{R}^n$ the gauge group action is always homotopically trivial, whereas the $\mathbb{T}^n$-theory is in general non-trivial.

Throughout, we studied observables in classical field theories that emerge as functions on the critical points of an action functional. However, we neglected so far that the derived critical locus is equipped with an extra geometrical structure, a *symplectic form*. Or put another way, following K. Costello and O. Gwilliams [CG16], a classical field theory is *defined* as a derived moduli problem possessing a symplectic form of cohomological degree -1. They also show that the classical factorization algebra of such a field theory carries a shifted Poisson structure. It is this shifted Poisson structure on the commutative factorization algebra of classical observables that allows to deform it, in a homotopical sense, into a factorization algebra of *quantum* observables. Hence, it is natural that the next step will be to study the compatibility between the shifted Poisson structure and the homotopy action we constructed.

Bibliography

[Bau14] Helga Baum. *Eichfeldtheorie - Eine Einführung in die Differentialgeometrie auf Faserbündeln.* Springer Spektrum, 2014.

[CG16] Kevin Costello and Owen Gwilliam. *Factorization algebras in quantum field theory*, volume 1 and 2. Cambridge University Press, 2016.

[CS74] Shiing-Shen Chern and James H. Simons. Characteristic forms and geometric invariants. *Ann. Math. (2)*, 99(1):48–69, 1974.

[DS95] William G. Dwyer and Jan Spalinski. Homotopy theories and model categories. In I. M. James, editor, *Handbook of algebraic topology*, pages 73–126. Elsevier Science, 1995.

[FHM05] Andrea Fuster, Marc Henneaux, and Axel Maas. BRST quantization: a short review. *Int. J. Geom. Meth. Mod. Phys.*, 2:939–964, 2005.

[Fre95] Daniel S. Freed. Classical Chern-Simons theory, part 1. *Adv. Math.*, 113:237–303, 1995.

[Fre02] Daniel S. Freed. Classical Chern-Simons theory, part 2. *Houston J. Math.*, 28(2):293–310, 2002.

[Fri12] Greg Friedmann. An elementary illustrated introduction to simplicial sets. *Rocky Mountain J. Math.*, 42(2):353–423, 2012.

[Get09] Ezra Getzler. Lie theory for nilpotent L_∞-algebras. *Ann. Math. (2)*, 170(1):271–301, 2009.

[GJ99] Paul G. Groess and John F. Jardine. *Simplicial Homotopy Theory.* Birkhäuser Verlag, 1999.

[Gwi12] Owen Gwilliam. *Factorization algebras and free field theories.* PhD thesis, Northwestern University, 2012.

[HS97] Peter J. Hilton and Urs Stammbach. *A course in homological algebra.* Springer, 1997.

C. Keller, *Chern-Simons Theory and Equivariant Factorization Algebras*, BestMasters, https://doi.org/10.1007/978-3-658-25338-7

[Ish13] Chris J. Isham. *Modern differential geometry for physicist.* World Scientific Publishing, 2013.

[KN96] Shoshichi Kobayashi and Katsumi Nomizu. *Foundations of differential geometry*, volume 1. Wiley, 1996.

[Lei14] Tom Leinster. *Basic category theory.* Cambridge University Press, 2014.

[Lur11] Jacob Lurie. Derived algebraic geometry X : formal moduli problems. 2011.

[LV12] Jean-Louis Loday and Bruno Vallette. *Algebraic operads.* Springer, 2012.

[Man98] Mihaela Manoliu. Abelian Chern-Simons theory. *J. Math. Phys.*, 39:170–206, 1998.

[Man09] Marco Manetti. Differential graded Lie algebras and formal deformation theory. *Proc. Sympos. Pure Math.*, 80:785–810, 2009.

[MDZ07] Martin Markl, Martin Doubek, and Petr Zima. Deformation theory (lecture notes). *Arch. Math.*, 43(5):333–371, 2007.

[Sto14] Stephan Stolz. Lecture notes: Functorial field theories and factorization algebras, September 2014.

[Vez11] Gabriele Vezzosi. Derived critical loci I - basics. *ArXive e-prints*, 2011.

[Wei95] Charles A. Weibel. *An introduction to homological algebra.* Cambridge University Press, 1995.

[Wen95] Xiao-Gang Wen. Topological orders and edge excitations in fractional quantum Hall states. *Adv. Phys.*, 44:405–473, 1995.

[Wit89a] Edward Witten. 2+1 dimenional gravity as an exactly soluble system. *Nucl. Phys.*, B 311:46–78, 1989.

[Wit89b] Edward Witten. Quantum field theory and the Jones polynomial. *Comm. Math. Phys.*, 121(3):351–399, 1989.

[Wit16] Edward Witten. Three lectures on topological phases of matter. *La Rivista del Nuovo Cimento*, 39(7):313–370, 2016.

[Zee95] Anthony Zee. Quantum Hall fluids. In H. B. Greyer, editor, *Field theory, topology and condensed matter physics*, pages 99–153. Berlin Springer Verlag, 1995.

Appendices

A Category Theory

This appendix provides a brief overview on categories, functors between categories and on natural transformations of functors. Moreover, we recall the notion of limits and colimits in categories, as well as representables and adjunctions. The main purpose is to settle the terminology used in this thesis. The main reference for the following is the book of T. Leinster [Lei14].

A.1 Categories and Functors

A.1.1 Categories

Definition A.1. A category $\mathcal{C}$ consists of

- a collection $\mathrm{ob}(\mathcal{C})$ of *objects*;
- for each $X, Y \in \mathrm{ob}(\mathcal{C})$, a collection $\mathrm{Hom}_{\mathcal{C}}(X, Y)$ of *morphisms*, also called *arrows*, from X to Y,

such that

- each object $X \in \mathrm{ob}(\mathcal{C})$ has s specific *identity morphism*

$$1_X : X \to X;$$

- for each $X, Y, Z \in \mathrm{ob}(\mathcal{C})$ and for each pair of morphisms $f \in \mathrm{Hom}_{\mathcal{C}}(X, Y)$ and $g \in \mathrm{Hom}_{\mathcal{C}}(Y, Z)$, there is a function

$$\begin{aligned} \mathrm{Hom}_{\mathcal{C}}(X, Y) \times \mathrm{Hom}_{\mathcal{C}}(Y, Z) &\to \mathrm{Hom}_{\mathcal{C}}(X, Z) \\ (f, g) &\mapsto g \circ f, \end{aligned}$$

  called *composition*.

This data has to satisfy the following axioms.

C. Keller, *Chern-Simons Theory and Equivariant Factorization Algebras*, BestMasters, https://doi.org/10.1007/978-3-658-25338-7

- For any $f \in \mathrm{Hom}_{\mathcal{C}}(X,Y)$, we have

$$1_Y \circ f = f \circ 1_X = f.$$

- For each $f \in \mathrm{Hom}_{\mathcal{C}}(X,Y)$, $g \in \mathrm{Hom}_{\mathcal{C}}(Y,Z)$ and $h \in \mathrm{Hom}_{\mathcal{C}}(Z,U)$ we have

$$h \circ (g \circ f) = (h \circ g) \circ f.$$

Remark A.1. We often write

- $X \in \mathcal{C}$ to mean $X \in \mathrm{ob}(\mathcal{C})$;
- $f : X \to Y$ to mean $f \in \mathrm{Hom}_{\mathcal{C}}(X,Y)$, and we call X the *domain* and Y the *codomain*;
- $\mathrm{mor}(\mathcal{C})$ to mean the collection of all morphisms in $\mathcal{C}$;
- gf to mean $g \circ f$.

Examples of Categories

Example A.1. *(Categories of mathematical structures).*

- There is a category **Set** whose objects are sets and whose morphisms are functions between sets.
- **Top** with topological spaces as objects and continuous functions as morphisms.
- **Grp** with groups as objects and homomorphisms of groups as morphisms.
- **Rings** with unital rings as objects and homomorphisms of rings as morphisms.
- $\mathbf{Vect}_k$ with vector spaces over a field k as objects and k-linear maps as morphisms.

Definition A.2. Let $\mathcal{C}$ be a category. The *opposite category* $\mathcal{C}^{\mathrm{op}}$ is the category defined by

- $\mathrm{ob}(\mathcal{C}^{\mathrm{op}}) \coloneqq \mathrm{ob}(\mathcal{C})$;
- $\mathrm{Hom}_{\mathcal{C}^{\mathrm{op}}}(X,Y) \coloneqq \mathrm{Hom}_{\mathcal{C}}(Y,X)$,

where

- for each $X \in \mathcal{C}$, 1_X^{op} is given by 1_X;

- for each $f^{\mathrm{op}} : X \to Y$ and $g^{\mathrm{op}} : Y \to Z$ the composition $g^{\mathrm{op}} \circ f^{\mathrm{op}}$ is given by $f \circ g$,

for all $X, Y, Z \in \mathcal{C}$.

Definition A.3. A *subcategory* $\mathcal{D} \subset \mathcal{C}$ is defined by restricting to a subcollection of objects and a subcollection of arrows satisfying the following.

- All arrows in $\mathcal{D}$ have domain and codomain in $\mathcal{D}$.
- Composition of all composable arrows in $\mathcal{D}$ lies in $\mathcal{D}$.
- The identity arrow lies in $\mathcal{D}$.

Definition A.4. A subcategory $\mathcal{D}$ of a category $\mathcal{C}$ is called *full* if

$$\mathrm{Hom}_{\mathcal{D}}(X, Y) = \mathrm{Hom}_{\mathcal{C}}(X, Y),$$

for all $X, Y \in \mathcal{D}$.

Types of Categories

Definition A.5. A category $\mathcal{C}$ is called *small* if the collections $\mathrm{ob}(\mathcal{C})$ and $\mathrm{mor}(\mathcal{C})$ are sets. Otherwise, $\mathcal{C}$ is called *large*.

Remark A.2. A category $\mathcal{C}$ such that $\mathrm{mor}(\mathcal{C})$ is a set is necessarily small.

Example A.2. Categories such as $\mathbf{Set}, \mathbf{Vect}_k, \mathbf{Grp}, \ldots$ are large.

Definition A.6. A category $\mathcal{C}$ is called *finite* if the collections $\mathrm{ob}(\mathcal{C})$ and $\mathrm{mor}(\mathcal{C})$ are finite sets.

Definition A.7. A category $\mathcal{C}$ is called *locally small* if for any X and Y in $\mathcal{C}$, the collection

$$\mathrm{Hom}_{\mathcal{C}}(X, Y)$$

is a set.

Remark A.3. In other words, a locally small category is a category enriched in $\mathbf{Set}$.

Example A.3. Categories such as $\mathbf{Set}, \mathbf{Vect}_k, \mathbf{Grp}, \ldots$ are locally small.

Types of Morphisms

Definition A.8. Let $\mathcal{C}$ be a category. A morphism $f : X \to Y$ in $\mathcal{C}$ is called an *isomorphism* if there exists a morphism $g : Y \to X$ such that

$$g \circ f = 1_X, \quad f \circ g = 1_Y.$$

For any $X, Y \in \mathcal{C}$ we say that X and Y are *isomorphic*, denoted $X \cong Y$, if there exists an isomorphism between X and Y.

Remark A.4. Given $f : X \to Y$ and $g : Y \to X$ such that $X \cong Y$, we call g an *inverse* of f and denote it by f^{-1}.

Definition A.9. A *groupoid* is a category $\mathcal{C}$ where all the arrows are invertible.

Example A.4. Let G be a group. We can associate to G a category BG in the following way

- $\text{ob}(BG) := \{\star\}$;
- $\text{mor}(BG) := \{G\}$.

In other words, any $g \in G$ can be regarded as an arrow $g : \star \to \star$, the unit element $e \in G$ is the identity arrow. Composition in BG is given by composition in G. Finally, any arrow in BG admits an inverse, namely the inverse of $g : \star \to \star$ is given by $g^{-1} : \star \to \star$.

Definition A.10. A morphism $f : X \to Y$ in a category $\mathcal{C}$ is called

- *monomorphism* if for all $h, g : Z \to X$ we have

$$f \circ h = f \circ g \implies h = g;$$

- *epimorphism* if for all $h, g : Y \to Z$ we have

$$h \circ f = g \circ f \implies h = g;$$

- *section* to an arrow $r : Y \to X$ if

$$r \circ f = 1_X;$$

- *retraction* to an arrow $s : Y \to X$ if

$$f \circ s = 1_Y.$$

Lemma A.1. *A morphism $f : X \to Y$ in $\mathcal{C}$ which is a section to $r : Y \to X$ is a monomorphism. A morphism $f : X \to Y$ in $\mathcal{C}$ which is a retraction to $s : Y \to X$ is an epimorphism.*

Remark A.5. A morphism which is a section is also called a *split monomorphism*. Dually, a morphism which is a retraction is also called a *split epimorphism*.

A.1.2 Functors

Functors encapsulate the notion of arrows between categories.

Definition A.11. Let $\mathcal{C}$ and $\mathcal{D}$ be categories. A *functor* $F : \mathcal{C} \to \mathcal{D}$ consists of the following data

- for each $X \in \mathcal{C}$ we have a function

$$\begin{aligned} \mathrm{ob}(\mathcal{C}) &\to \mathrm{ob}(\mathcal{D}) \\ X &\mapsto F(X), \end{aligned}$$

- for each $X, Y \in \mathcal{C}$ we have a function

$$\begin{aligned} \mathrm{Hom}_{\mathcal{C}}(X,Y) &\to \mathrm{Hom}_{\mathcal{D}}(F(X),F(Y)) \\ f &\mapsto F(f), \end{aligned}$$

subjected to the following axioms

$$F(1_X) = 1_{F(X)}, \quad F(g \circ f) = F(g) \circ F(f),$$

for all $f \in \mathrm{Hom}_{\mathcal{C}}(X,Y)$, $g \in \mathrm{Hom}_{\mathcal{C}}(Y,Z)$ and $X, Y, Z \in \mathcal{C}$.

Definition A.12. A functor $F : \mathcal{C} \to \mathcal{D}$ is called *faithful* (respectively, *full*) if for all $X, Y \in \mathcal{C}$ the following assignment is injective (respectively, surjective).

$$\begin{aligned} \mathrm{Hom}_{\mathcal{C}}(X,Y) &\to \mathrm{Hom}_{\mathcal{D}}(F(X),F(Y)) \\ f &\mapsto F(f) \end{aligned}$$

Examples of Functors

Example A.5. *(Forgetful functors).*

- There is a functor $U : \mathbf{Grp} \to \mathbf{Set}$ defined as follows. If G is a group then $U(G)$ is its underlying set of elements and if $f : G \to H$ is a group homomorphism then $U(f) : U(G) \to U(H)$ is the function between the underlying sets, that is the functor U forgets the group structure of groups and forgets that group homomorphisms are homomorphisms.
- Similarly, for any field k there is a functor $U : \mathbf{Vect}_k \to \mathbf{Set}$ forgetting the vector space structure.
- There is a functor $U : \mathbf{Rings} \to \mathbf{Ab}$ that forgets the multiplicative structure, remembering only the underlying abelian group.

Example A.6. *(Free functors).*

- There is a functor $F : \mathbf{Set} \to \mathbf{Grp}$ defined on objects as the free group $F(S)$ over the set S. Recall that the free group $F(S)$ consists of all expressions that can be built from elements in S. The group structure on $F(S)$ is given by concatenation.
- For a fixed field k, there is a functor $F : \mathbf{Set} \to \mathbf{Vect}_k$ defined on objects by taking $F(S)$ to be the free k-vector space with basis S. Broadly speaking, $F(S)$ is the set of all formal k-linear combinations of elements of S, that is, expressions of the form

$$\sum_{s \in S} \lambda_s s$$

where $\lambda_s \in k$ and there are only finitely many s such that $\lambda_s \neq 0$. Addition in $F(S)$ is given by

$$\Big(\sum_{s \in S} \lambda_s s\Big) + \Big(\sum_{s \in S} \bar{\lambda}_s s\Big) = \sum_{s \in S} (\lambda_s + \bar{\lambda}_s) s$$

and scalar multiplication on $F(S)$ by

$$c \cdot \sum_{s \in S} \lambda_s s = \sum_{s \in S} (c\lambda_s) s$$

for $c \in k$. Hence, $F(S)$ is a k-vector space.

Example A.7. Let G be a group regarded as the one-object category BG. Then, a functor $\varphi : BG \to \mathbf{Set}$ is a set equipped with a left G-action. Indeed, the functor φ consists of a set S together with functions $\varphi(g) : S \to S$ for every $g \in G$. Writing $\varphi(g)(s) := g.s$, we see that φ is a set S together with a function

$$G \times S \to S$$
$$(g, s) \mapsto g.s,$$

satisfying $(gg').s = g.(g'.s)$ and $e.s = s$ for all $g, g' \in G$ and $s \in S$. Hence, φ is a left G-set.

Definition A.13. Let $\mathcal{C}$ and $\mathcal{D}$ be categories. A *contravariant functor* from $\mathcal{C}$ to $\mathcal{D}$ is a functor $\mathcal{C}^{\mathrm{op}} \to \mathcal{D}$.

In chapter 5 we introduce the notion of *presheaves* on topological spaces. Recall that a presheaf F on a topological space M assigns to every open subset $U \subset M$ a set $F(U)$ and to any inclusion $U \hookrightarrow V$ of open subsets a restriction map $F(V) \to F(U)$. If we write $\mathbf{Opens}(M)$ for the category whose objects are open subsets of M and whose morphism are given by inclusion, meaning that there is a morphism from U to V exactly if $U \subset V$, then a presheaf is simply a contravariant functor $F : \mathbf{Opens}(M)^{\mathrm{op}} \to \mathbf{Set}$. Or more generally, we define a presheaf on an arbitrary category as follows.

Definition A.14. Let $\mathcal{C}$ be a category. A *presheaf* on $\mathcal{C}$ is a functor

$$\mathcal{C}^{\mathrm{op}} \to \mathbf{Set}.$$

A.2 Natural Transformations

Suppose we have given two functors $F, G : \mathcal{C} \to \mathcal{D}$ with the same domain and codomain. Then, there is the notion of a map between F and G. Those maps are called *natural transformations.*

Definition A.15. Let $\mathcal{C}$ and $\mathcal{D}$ be categories and let $F, G : \mathcal{C} \to \mathcal{D}$ be functors. A *natural transformation* α from F to G is a family

$$\left(F(X) \xrightarrow{\alpha_X} G(X)\right)_{X \in \mathcal{C}}$$

of maps in $\mathcal{D}$, such that for any map $f : X \to Y$ in $\mathcal{C}$ the following diagram

$$\begin{array}{ccc} F(X) & \xrightarrow{F(f)} & F(Y) \\ {\scriptstyle \alpha_X}\downarrow & & \downarrow{\scriptstyle \alpha_Y} \\ G(X) & \xrightarrow[G(f)]{} & G(Y) \end{array}$$

is commutative for all $X, Y \in \mathcal{C}$. The maps α_X are called *components* of α.

Remark A.6. It is common to write

$$\mathcal{C} \underset{G}{\overset{F}{\underset{\Downarrow \alpha}{\rightrightarrows}}} \mathcal{D}$$

or simply $\alpha : F \Rightarrow G$ to mean that α is a natural transformation from F to G.

Example A.8. Recall from example A.7 that a functor $BG \to \mathbf{Set}$ is nothing but a G-set S. Given two G-sets S and T, a natural transformation $S \Rightarrow T$ is a function of sets $\alpha : S \to T$, such that $\alpha(g.s) = g.\alpha(s)$ for all $g \in G$ and $s \in S$. Hence, α is a G-equivariant map.

Functor Categories

We can compose natural transformations. Namely, given functors $F, G, H : \mathcal{C} \to \mathcal{D}$ and natural transformations $\alpha : F \Rightarrow G$ and $\beta : G \Rightarrow H$, there is a composite $\beta \circ \alpha : F \Rightarrow H$ defined by

$$(\beta \circ \alpha)_X = \beta_X \circ \alpha_X,$$

for all $X \in \mathcal{C}$. Moreover, there is a natural identity transformation $1_F : F \Rightarrow F$ on any functor $F : \mathcal{C} \to \mathcal{D}$ defined by $(1_F)_X = 1_{F(X)}$ for all $X \in \mathcal{C}$. Hence, we have the following definition.

Definition A.16. Let $\mathcal{C}$ and $\mathcal{D}$ be categories. The *functor category* $[\mathcal{C}, \mathcal{D}]$ is the category whose

- objects are functors $F : \mathcal{C} \to \mathcal{D}$;
- morphisms are natural transformations $\alpha : F \Rightarrow G$.

Definition A.17. Let $\mathcal{C}$ and $\mathcal{D}$ be categories. A *natural isomorphism* between functors from $\mathcal{C}$ to $\mathcal{D}$ is an isomorphism in $[\mathcal{C}, \mathcal{D}]$.

Lemma A.2. *Let $F, G : \mathcal{C} \to \mathcal{D}$ be functors and $\alpha : F \Rightarrow G$ a natural transformation between them. Then α is a natural isomorphism if and only if $\alpha_X : F(X) \to G(X)$ is an isomorphism for all $X \in \mathcal{C}$.*

Definition A.18. Given functors $F, G : \mathcal{C} \to \mathcal{D}$, we say that $F(X) \cong G(X)$ *naturally* in $X \in \mathcal{C}$, if F and G are naturally isomorphic.

The appropriate notion of 'sameness' for two functors is thus being naturally isomorphic. The following definition gives the corresponding notion for categories.

Definition A.19. An *equivalence* between categories $\mathcal{C}$ and $\mathcal{D}$ consists of a pair of functors $F : \mathcal{C} \leftrightarrows \mathcal{D} : G$ together with natural isomorphisms

$$\eta : 1_{\mathcal{C}} \to G \circ F \quad \text{and} \quad \epsilon : F \circ G \to 1_{\mathcal{D}}.$$

Remark A.7. If there exists an equivalence between $\mathcal{C}$ and $\mathcal{D}$, we say that $\mathcal{C}$ and $\mathcal{D}$ are *equivalent* and we write $\mathcal{C} \simeq \mathcal{D}$. The functors F and G are called *equivalences*.

We want to give a characterization of those functors that are equivalences. To that end, we need the following definition.

Definition A.20. A functor $F : \mathcal{C} \to \mathcal{D}$ is called *essentially surjective on objects* if for all $Y \in \mathcal{D}$ there exists an $X \in \mathcal{C}$ such that $F(X) \cong Y$.

Theorem A.1. *A functor is an equivalence if and only if it is full, faithful and essentially surjective on objects.*

A.3 Limits

We first give the definition of a *limit* in a category, which is very general. We also introduce the dual notion of a *colimit*. We then examine some particularly useful types of (co-)limits, such as (co-)products, (co-)equalizers, pullbacks and pushouts respectively.

A.3.1 Definition of Limits

Definition A.21. Let $\mathcal{C}$ be a category and J a small category. A functor $D : J \to \mathcal{C}$ is called a *diagram* in $\mathcal{C}$ of shape J.

Definition A.22. Let $\mathcal{C}$ be a category and $D : J \to \mathcal{C}$ a diagram. A *cone* on D is an object $X \in \mathcal{C}$, called the *vertex*, together with a family

$$\left(X \xrightarrow{f_j} D(j)\right)_{j \in J}$$

of morphisms in $\mathcal{C}$, such that for any morphism $u : j \to j'$ in J the following diagram

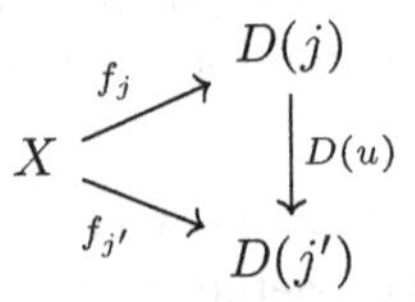

$$\begin{array}{ccc} & & D(j) \\ & \overset{f_j}{\nearrow} & \\ X & & \downarrow D(u) \\ & \underset{f_{j'}}{\searrow} & \\ & & D(j') \end{array}$$

is commutative.

Definition A.23. A *limit* of a diagram $D : J \to \mathcal{C}$ is a cone on D

$$\left(L \xrightarrow{p_j} D(j)\right)_{j \in J}$$

such that for any other cone $(X \xrightarrow{f_j} D(j))_{j \in J}$ on D there exists a unique morphism $\bar{f} : X \to L$ such that $p_j \circ \bar{f} = f_j$ for all $j \in J$. The maps p_j are called the *projections* of the limit.

Remark A.8. Sometimes we refer to L as the limit cone and write

$$L = \lim_{\leftarrow J} D.$$

Existence of Limits

Definition A.24. Let J be a small category. A category $\mathcal{C}$ *has limits of shape* J if for every diagram D of shape J in $\mathcal{C}$, a limit of D exists.

Definition A.25. A category *has all limits* (respectively, *has finite limits*) if it has limits of shape J for all small (respectively, finite) categories J.

Remark A.9. We say a category $\mathcal{C}$ is *complete* if it has all limits.

Example A.9. Categories such as $\mathbf{Set}, \mathbf{Vect}_k, \mathbf{Grp}, \mathbf{Top}, \ldots$ are all complete.

Lemma A.3. *Let $\mathcal{C}$ be a category.*

- *If $\mathcal{C}$ has all products and equalizers then $\mathcal{C}$ has all limits.*
- *If $\mathcal{C}$ has binary products, a terminal object and equalizers then $\mathcal{C}$ has finite limits.*

Limits as Terminal Objects

Definition A.26. Let $\mathcal{C}$ be a category. An object $T \in \mathcal{C}$ is called *terminal* if for every $X \in \mathcal{C}$, there is exactly one morphism $X \to T$.

Definition A.27. Let $\mathcal{C}$ and J be a categories. The *constant functor*

$$\Delta X : J \to \mathcal{C}$$

is defined for any object $X \in \mathcal{C}$ as follows

- on objects by $\Delta X(j) \coloneqq X$;
- on morphisms $u : j \to j'$ by $\Delta X(u) \coloneqq 1_X$,

for all $j, j' \in J$.

The constant functors given by the objects of a category $\mathcal{C}$ assemble into the so-called *diagonal functor* of $\mathcal{C}$.

Definition A.28. Let $\mathcal{C}$ be a category and J a small category. The *diagonal functor*

$$\Delta : \mathcal{C} \to [J, \mathcal{C}]$$

is defined

- on objects $X \in \mathcal{C}$ by $\Delta(X) \coloneqq \Delta X$, where ΔX is the constant functor;
- on morphisms $f : X \to Y$ as the natural transformation $\Delta(f) : \Delta X \Rightarrow \Delta Y$. But since ΔX and ΔY are constant functors, $\Delta(f)$ is simply the morphism $f : X \to Y$ for all objects in J.

Now let $X \in \mathcal{C}$ and let $D : J \to \mathcal{C}$ be a diagram. A natural transformation $\alpha : \Delta X \Rightarrow D$ is the same thing as a cone on D. Indeed, the components of the natural transformations α are given by

$$\left(X \xrightarrow{\alpha_j} D(j)\right)_{j \in J},$$

such that for all morphisms $u : j \to j'$ the following diagram is commutative

$$\begin{array}{ccc} \Delta X(j) = X & \xrightarrow{\alpha_j} & D(j) \\ {\scriptstyle \Delta X(u) = 1_X}\downarrow & & \downarrow{\scriptstyle D(u)} \\ \Delta X(j') = X & \xrightarrow{\alpha_{j'}} & D(j') \end{array}$$

Definition A.29. Let $D : J \to \mathcal{C}$ be a diagram. Denote with $\mathbf{Cone}(D)$ the category whose

- objects (X, α) are natural transformation $\alpha : \Delta X \Rightarrow D$, for all $X \in \mathcal{C}$;
- morphisms $(X, \alpha) \to (X', \beta)$ are maps $f : X \to X'$ in $\mathcal{C}$ such that the diagram

$$\begin{array}{ccc} \Delta X & \xRightarrow{\Delta(f)} & \Delta X' \\ & {\scriptstyle \alpha}\searrow \quad \swarrow{\scriptstyle \beta} & \\ & D & \end{array}$$

is commutative.

Lemma A.4. *A* limit *of a diagram $D : J \to \mathcal{C}$ is the terminal object in* $\mathbf{Cone}(D)$.

A.3.2 Products, Equalizers, Pullbacks

Let us give some concrete examples of limits in a category $\mathcal{C}$.

Products

Let $\mathbf{I}$ be the small category with two objects. Then, a diagram $D : \mathbf{I} \to \mathcal{C}$ is a pair (X_1, X_2) of objects in $\mathcal{C}$.

$$\mathbf{I} \quad = \quad \bullet \quad \bullet$$

Definition A.30. A *product* $P = X_1 \times X_2$ is a limit cone on $D : \mathbf{I} \to \mathcal{C}$, that is a cone

$$(P \xrightarrow{p_i} X_i)_{i=1,2}$$

such that for any other cone $(A \xrightarrow{f_i} X_i)_{i=1,2}$ the diagram

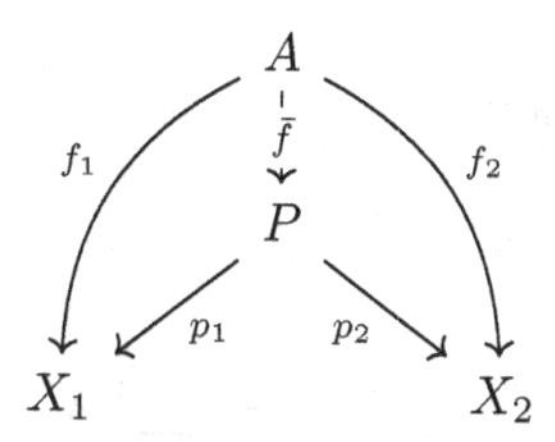

is commutative.

Remark A.10. Similarly, we can define the product of a family of objects $\{X_j\}_{j \in J}$. If the product exists, it is common to write $P = \prod_{j \in J} X_j$.

Example A.10. Any two sets X and Y have a product in **Set**, given by the usual Cartesian product $X \times Y$, equipped with the usual projection maps p_1 and p_2.

Equalizers

We first need the following definition.

Definition A.31. Let A, X, Y be objects in $\mathcal{C}$. A *fork* in $\mathcal{C}$ consists of objects and morphisms

$$A \xrightarrow{f} X \overset{s}{\underset{t}{\rightrightarrows}} Y$$

such that $s \circ f = t \circ f$.

Now, let $\mathbf{J}$ be the small category with two objects and two non-identity arrows.

$$\mathbf{J} = \bullet \rightrightarrows \bullet$$

Then, a diagram $D : \mathbf{J} \to \mathcal{C}$ is a pair

$$X \underset{t}{\overset{s}{\rightrightarrows}} Y$$

of morphisms in $\mathcal{C}$. A cone on D consists of objects and morphisms

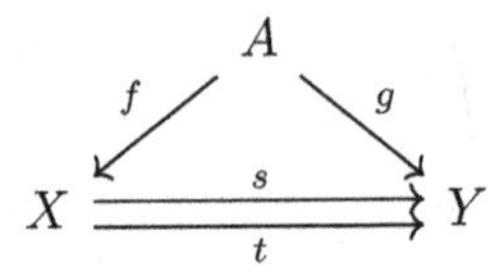

such that $s \circ f = g$ and $t \circ f = g$. But since g is determined by f, we can say that a cone on D consists of an object A and a morphism $f : A \to X$ such that

$$A \xrightarrow{f} X \underset{t}{\overset{s}{\rightrightarrows}} Y$$

is a fork.

Definition A.32. An *equalizer* E is a limit cone on $D : \mathbf{J} \to \mathcal{C}$, that is an object $E \in \mathcal{C}$ and a morphism $i : E \to X$ such that

$$E \xrightarrow{i} X \underset{t}{\overset{s}{\rightrightarrows}} Y$$

is a fork, and given any other cone (A, f) on D there exists a unique map $\bar{f} : A \to E$ such that the follwoing diagram is commutative.

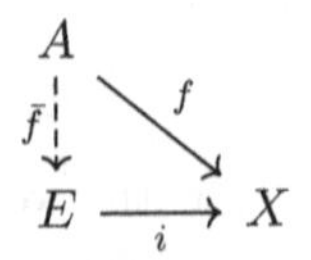

Example A.11. Take objects and morphisms

$$X \underset{t}{\overset{s}{\rightrightarrows}} Y$$

in **Set**. The equalizer is the set

$$\{x \in E \mid s(x) = t(x)\},$$

together with the inclusion $i : E \hookrightarrow X$ of sets. Then $s \circ i = t \circ i$ and one can check that (E, i) is the universal fork on s and t.

Pullbacks

Let $\mathbf{K}$ be the small category

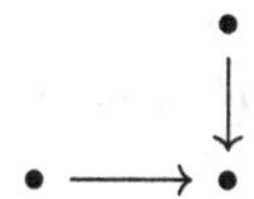

Then, a diagram $D : \mathbf{K} \to \mathcal{C}$ consists of objects and morphisms

$$\begin{array}{ccc} & & Y \\ & & \downarrow{\scriptstyle t} \\ X & \xrightarrow[s]{} & Z \end{array}$$

in $\mathcal{C}$. Performing a similar simplification as in the previous section, we find that a cone on D is a commutative square

$$\begin{array}{ccc} A & \xrightarrow{f_2} & Y \\ {\scriptstyle f_1}\downarrow & & \downarrow{\scriptstyle t} \\ X & \xrightarrow[s]{} & Z \end{array}$$

in $\mathcal{C}$.

Definition A.33. A *pullback* $P = X \times_Z Y$, also called *fiber product*, is a limit cone on $D : \mathbf{K} \to \mathcal{C}$, that is an object $P \in \mathcal{C}$ together with morphisms $p_1 : P \to X$ and $p_2 : P \to Y$ such that the diagram

$$\begin{array}{ccc} P & \xrightarrow{p_2} & Y \\ {\scriptstyle p_1}\big\downarrow & & \big\downarrow{\scriptstyle t} \\ X & \xrightarrow[s]{} & Z \end{array}$$

is commutative, and with the property that for any other cone on D, there is a unique morphism $\bar{f} : A \to P$ such that

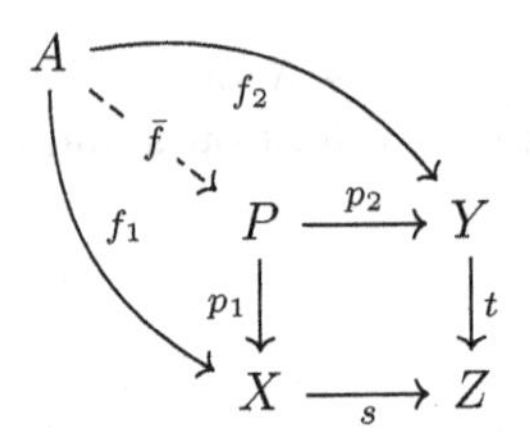

is commutative.

Remark A.11. If the pullback P exists it is common to denote the diagram by

$$\begin{array}{ccc} P \lrcorner & \xrightarrow{p_2} & Y \\ {\scriptstyle p_1}\big\downarrow & & \big\downarrow{\scriptstyle t} \\ X & \xrightarrow[s]{} & Z \end{array}$$

Example A.12. The pullback in **Set** is

$$P = \{(x, y) \in X \times Y \mid s(x) = t(y)\}$$

with projections p_1 and p_2 given by $p_1(x, y) = x$ and $p_2(x, y) = y$.

A.3.3 Definition of Colimits

We dualize the notion of a limit in a category.

Definition A.34. Let $\mathcal{C}$ be a category, J a small category and $D : J \to \mathcal{C}$ a diagram. A *cocone* on D is an object $X \in \mathcal{C}$, called the *vertex*, together with a family

$$\left(D(j) \xrightarrow{f_j} X\right)_{j \in J}$$

of morphisms in $\mathcal{C}$, such that for any morphism $u : j \to j'$ in J the following diagram

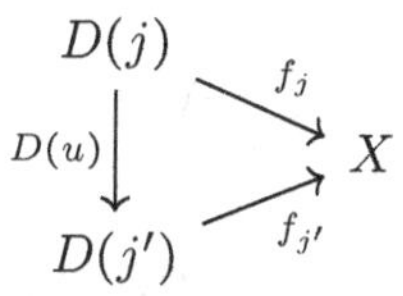

$$\begin{array}{ccc} D(j) & \searrow^{f_j} & \\ {\scriptstyle D(u)}\downarrow & & X \\ D(j') & \nearrow_{f_{j'}} & \end{array}$$

is commutative.

Definition A.35. A *colimit* of a diagram $D : J \to \mathcal{C}$ is a cocone on D

$$\left(D(j) \xrightarrow{p_j} L\right)_{j \in J},$$

such that for any other cocone $(D(j) \xrightarrow{f_j} X)_{j \in J}$ on D there exists a unique morphism $\bar{f} : L \to X$ such that $\bar{f} \circ p_j = f_j$ for all $j \in J$. The maps p_j are called the *coprojections* of the colimit.

Colimits as Initial Objects

Definition A.36. Let $\mathcal{C}$ be a category. An object $I \in \mathcal{C}$ is called *initial* if for every $X \in \mathcal{C}$, there is exactly one morphism $I \to X$.

Let $\mathcal{C}$ be a category, J a small category and $D : J \to \mathcal{C}$ a diagram. A natural transformation $\alpha : D \Rightarrow \Delta X$ is the same thing as a cocone on D. Here, ΔX is the constant functor from definition A.27.

Definition A.37. Let $D : J \to \mathcal{C}$ be a diagram. Denote with $\mathbf{Cocone}(D)$ the category whose

- objects (X, α) are natural transformations $\alpha : D \Rightarrow \Delta X$, for all $X \in \mathcal{C}$;
- morphisms $(X, \alpha) \to (X', \beta)$ are maps $f : X \to X'$ in $\mathcal{C}$ such that the following diagram is commutative.

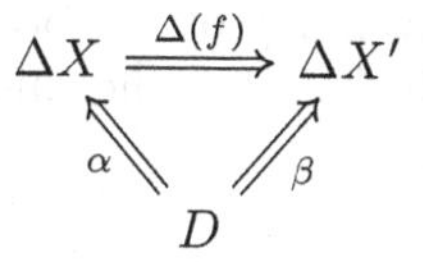

$$\begin{array}{ccc} \Delta X & \xRightarrow{\Delta(f)} & \Delta X' \\ {\scriptstyle \alpha}\Uparrow\!\!\!\!\!\nwarrow & & \nearrow\!\!\!\!\!\Uparrow{\scriptstyle \beta} \\ & D & \end{array}$$

Lemma A.5. *A* colimit *of a diagram $D : J \to \mathcal{C}$ is the initial object in* $\mathbf{Cocone}(D)$.

A.3.4 Coproducts, Coequalizers, Pushouts

Coproducts

Definition A.38. A *coproduct* $P = X_1 + X_2$, also called *sum*, is a colimit on $D : \mathbf{I} \to \mathcal{C}$, that is a cocone

$$\left(X_i \xrightarrow{p_i} P\right)_{i=1,2},$$

such that for any other cocone $(X_i \xrightarrow{f_i} A)_{i=1,2}$ the diagram

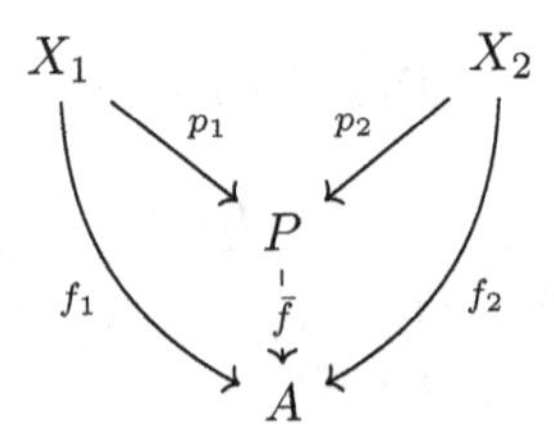

is commutative.

Example A.13. Any two sets X and Y have a coproduct in **Set** given by the disjoint union

$$P = X \sqcup Y$$

and the coprojections are given by inclusion.

Coequalizers

Definition A.39. A *coequalizer* E is a colimit on $D : \mathbf{J} \to \mathcal{C}$, that is an object $E \in \mathcal{C}$ and a morphism $e : Y \to E$ such that in the diagram

$$X \underset{t}{\overset{s}{\rightrightarrows}} Y \xrightarrow{\ e\ } E$$

we have $e \circ s = e \circ t$, and given any other cocone (A, f) on D there exists a unique map $\bar{f} : E \to A$ such that the following diagram is commutative.

$$\begin{array}{ccc} & & A \\ & \overset{f}{\nearrow} & \uparrow \bar{f} \\ Y & \xrightarrow[e]{} & E \end{array}$$

Example A.14. Take objects and morphisms

$$X \underset{t}{\overset{s}{\rightrightarrows}} Y$$

in **Set**. The coequalizer is the universal set E together with a function $e : Y \to E$ such that $e(s(x)) = e(t(x))$ for all $x \in X$. We have to canonically construct E. To that end, let $\sim$ be the equivalence relation on Y generated by $s(x) \sim t(x)$ for all $x \in X$. Define $E := Y/\sim$ and take e to be the quotient map $e : Y \to E$.

Pushouts

Let $\mathbf{K}^{\mathrm{op}}$ be the small category

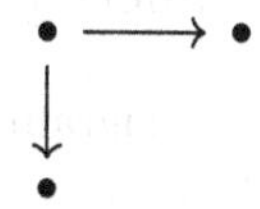

Then, a diagram $D^{\mathrm{op}} : \mathbf{K}^{\mathrm{op}} \to \mathcal{C}^{\mathrm{op}}$ consists of objects and morphisms

$$\begin{array}{ccc} Z & \xrightarrow{t} & Y \\ {\scriptstyle s}\downarrow & & \\ X & & \end{array}$$

Definition A.40. A *pushout* $P = X +_Z Y$ is a colimit of $D^{\mathrm{op}} : \mathbf{K}^{\mathrm{op}} \to \mathcal{C}^{\mathrm{op}}$, that is an object $P \in \mathcal{C}$ together with morphisms $p_1 : X \to P$ and $p_2 : Y \to P$ such that the diagram

$$\begin{array}{ccc} Z & \xrightarrow{t} & Y \\ {\scriptstyle s}\downarrow & & \downarrow{\scriptstyle p_2} \\ X & \xrightarrow[p_1]{} & P \end{array}$$

is commutative, and with the property that for any other cocone on D^{op}, there is a unique morphism $\bar{f} : P \to A$ such that

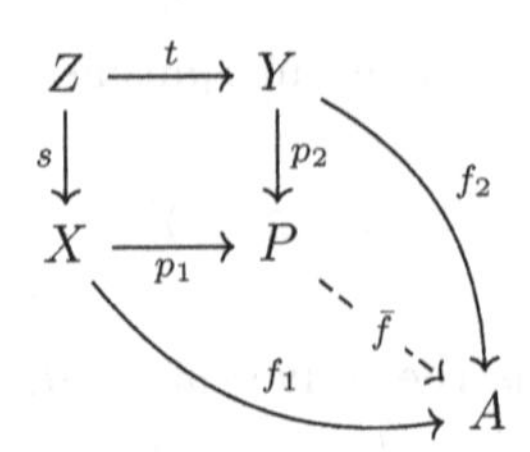

is commutative.

Example A.15. The pushout P in **Set** is $(X \sqcup Y)/\sim$, where $\sim$ is the equivalence relation on $X \sqcup Y$ generated by $s(z) \sim t(z)$ for all $z \in Z$. The coprojection $p_1 : X \to P$ sends $x \in X$ to its equivalence class in P, and similarly for the coprojection $p_2 : Y \to P$.

A.3.5 Interactions between Functors and Limits

Let $\mathcal{C}$ be a category and $D : J \to \mathcal{C}$ a diagram in $\mathcal{C}$. Given a functor $F : \mathcal{C} \to \mathcal{D}$, we get a diagram $F \circ D : J \to \mathcal{D}$ in $\mathcal{D}$.

Definition A.41. A functor $F : \mathcal{C} \to \mathcal{D}$ is said to *preserve limits of shape* J, if for all diagrams $D : J \to \mathcal{C}$ with a limit cone we have

$$\left(L \xrightarrow{p_j} D(j)\right)_{j\in J} \text{ is a limit cone on } D \Rightarrow \left(F(L) \xrightarrow{F(p_j)} FD(j)\right)_{j\in J} \text{ is a limit cone on } F \circ D.$$

Definition A.42. A functor $F : \mathcal{C} \to \mathcal{D}$ is said to *reflect limits of shape* J, if for all diagrams $D : J \to \mathcal{C}$ and cones on D we have

$$\left(F(L) \xrightarrow{F(p_j)} FD(j)\right)_{j\in J} \text{ is a limit cone on } F \circ D \Rightarrow \left(L \xrightarrow{p_j} D(j)\right)_{j\in J} \text{ is a limit cone on } D.$$

Definition A.43. A functor $F : \mathcal{C} \to \mathcal{D}$ is said to *create limits of shape* J, if whenever $D : J \to \mathcal{C}$ is a diagram in $\mathcal{C}$ we have

- for any limit cone $(L \xrightarrow{p_j} FD(j))_{j\in J}$ on the diagram $F \circ D$, there is a unique cone $(\bar{L} \xrightarrow{\bar{p}_j} D(j))_{j\in J}$ on D such that $F(\bar{L}) = L$ and $F(\bar{p}_j) = p_j$ for all $j \in J$;

– this cone $(\bar{L} \xrightarrow{\bar{p}_j} D(j))_{j \in J}$ is a limit cone on D.

Remark A.12. The same terminology applies to colimits.

A.4 Representables

In this section we address the question of how a given object 'sees' the rest of the category it lives in. In other words, we are going to consider the totality of maps out of a fixed object. For instance, take the one-point space $1 \in \mathbf{Top}$. A morphism from 1 to any other topological space X is essentially the same as a point of X. Similarly, a morphism from an interval $I \subset \mathbb{R}$ to X is a curve in X. Hence, 1 'sees' points, whereas I 'sees' curves. Formalizing these observations leads to the notion of *representable* functors.

Definition A.44. Let $\mathcal{C}$ be a locally small category and let $X \in \mathcal{C}$. We have a functor

$$H^X = \mathrm{Hom}_{\mathcal{C}}(X, -) : \mathcal{C} \to \mathbf{Set},$$

defined as follows

– on objects $Y \in \mathcal{C}$ by

$$H^X(Y) := \mathrm{Hom}_{\mathcal{C}}(X, Y);$$

– on morphisms $f : Y \to Y'$ in $\mathcal{C}$ by

$$\begin{aligned} H^X(f) : \mathrm{Hom}_{\mathcal{C}}(X, Y) &\to \mathrm{Hom}_{\mathcal{C}}(X, Y') \\ g &\mapsto f \circ g \end{aligned}$$

Remark A.13. Sometimes $H^X(f)$ is written as f_* or $\mathrm{Hom}_{\mathcal{C}}(X, f)$.

Definition A.45. Let $\mathcal{C}$ be a locally small category. A functor $F : \mathcal{C} \to \mathbf{Set}$ is called *representable* if

$$F \cong H^X$$

for some $X \in \mathcal{C}$. A *representation* of F is a choice of an object $X \in \mathcal{C}$ and of a natural isomorphism between H^X and F.

Example A.16. Consider $H^1 : \mathbf{Set} \to \mathbf{Set}$, where $1 = \{\star\}$ is the 1-element set. Since $\mathrm{Hom}_{\mathbf{Set}}(1, A) \cong A$, we have

$$H^1(A) \cong A$$

for each $A \in \mathbf{Set}$. This isomorphism is natural in A. So we have a natural isomorphism

$$H^1 \cong 1_{\mathbf{Set}}$$

Hence, $1_{\mathbf{Set}}$ is representable.

For each object $X \in \mathcal{C}$, definition A.44 yields a functor $H^X \in [\mathcal{C}, \mathbf{Set}]$. We expect that if we vary X in $\mathcal{C}$, the functors H^X should vary in $[\mathcal{C}, \mathbf{Set}]$ in a consistent way and thus patch together somehow to a family $\{H^X\}_{X \in \mathcal{C}}$. Precisely, a morphism $f : X \to Y$ in $\mathcal{C}$ induces a natural transformation

$$H^f : H^Y \Rightarrow H^X,$$

whose components are defined by

$$\begin{aligned} H^f_Z : H^Y(Z) = \mathrm{Hom}_{\mathcal{C}}(Y, Z) &\to H^X(Z) = \mathrm{Hom}_{\mathcal{C}}(X, Z) \\ g &\mapsto g \circ f, \end{aligned}$$

for all $Z \in \mathcal{C}$.

Definition A.46. Let $\mathcal{C}$ be a locally small category. The functor

$$H^\bullet : \mathcal{C}^{\mathrm{op}} \to [\mathcal{C}, \mathbf{Set}]$$

is defined on objects X by $H^\bullet(X) \coloneqq H^X$ and on morphisms f by $H^\bullet(f) \coloneqq H^f$.

We now dualize the definitions presented so far. In words, instead of considering the maps out of a given objects we now consider the totality of maps into a given object, answering the question how an object is seen from the rest of the category.

Definition A.47. Let $\mathcal{C}$ be a locally small category and let $X \in \mathcal{C}$. We have a functor

$$H_X = \mathrm{Hom}_{\mathcal{C}}(-, X) : \mathcal{C}^{\mathrm{op}} \to \mathbf{Set},$$

defined as follows

- on objects $Y \in \mathcal{C}$ by

$$H_X(Y) \coloneqq \mathrm{Hom}_{\mathcal{C}}(Y, X);$$

– on morphisms $f : Y \to Y'$ in $\mathcal{C}$ by

$$\begin{aligned} H_X(f) : \mathrm{Hom}_{\mathcal{C}}(Y', X) &\to \mathrm{Hom}_{\mathcal{C}}(Y, X) \\ g &\mapsto g \circ f \end{aligned}$$

Definition A.48. Let $\mathcal{C}$ be a locally small category. A functor $F : \mathcal{C}^{\mathrm{op}} \to \mathbf{Set}$ is called *corepresentable* if

$$F \cong H_X$$

for some $X \in \mathcal{C}$. A *corepresentation* of F is a choice of an object $X \in \mathcal{C}$ and of a natural isomorphism between H_X and F.

Given a morphism $f : X \to Y$ in $\mathcal{C}$, we have a natural transformation

$$H_f : H_X \Rightarrow H_Y,$$

whose components are defined by

$$\begin{aligned} H_f(Z) : H_X(Z) = \mathrm{Hom}_{\mathcal{C}}(Z, X) &\to H_Y(Z) = \mathrm{Hom}_{\mathcal{C}}(Z, Y) \\ g &\mapsto f \circ g, \end{aligned}$$

for all $Z \in \mathcal{C}$.

Definition A.49. Let $\mathcal{C}$ be a locally small category. The *Yoneda embedding* of $\mathcal{C}$ is the functor

$$H_\bullet : \mathcal{C} \to [\mathcal{C}^{\mathrm{op}}, \mathbf{Set}],$$

defined on objects $X \in \mathcal{C}$ by $H_\bullet(X) := H_X$ and on morphisms f by $H_\bullet(f) := H_f$.

In summary, for any locally small category $\mathcal{C}$, we have the following.

- for each $X \in \mathcal{C}$ we have a functor $H^X : \mathcal{C} \to \mathbf{Set}$;
- the functors $\{H^X\}_{X \in \mathcal{C}}$ assemble in a functor $H^\bullet : \mathcal{C}^{\mathrm{op}} \to [\mathcal{C}, \mathbf{Set}]$;
- for each $X \in \mathcal{C}$ we have a functor $H_X : \mathcal{C}^{\mathrm{op}} \to \mathbf{Set}$;
- the functors $\{H_X\}_{X \in \mathcal{C}}$ assemble in a functor $H_\bullet : \mathcal{C} \to [\mathcal{C}^{\mathrm{op}}, \mathbf{Set}]$.

A.4.1 The Yoneda Lemma

Given a representable presheaf $H_X : \mathcal{C}^{\mathrm{op}} \to \mathbf{Set}$, the Yoneda lemma characterizes the maps from H_X into any other object in the presheaf category $[\mathcal{C}^{\mathrm{op}}, \mathbf{Set}]$. Informally, the Yoneda lemma states that for any $X \in \mathcal{C}$ and presheaf F on $\mathcal{C}$, a natural transformation $H_X \Rightarrow F$ is an element of $F(X)$.

Theorem A.2. *(Yoneda). Let $\mathcal{C}$ be a locally small category. Then*

$$\mathrm{Hom}_{[\mathcal{C}^{\mathrm{op}},\mathbf{Set}]}(H_X, F) \simeq F(X), \qquad (\star)$$

naturally in $X \in \mathcal{C}$ and $F \in [\mathcal{C}^{\mathrm{op}}, \mathbf{Set}]$.

Remark A.14. The natural isomorphism $(\star)$ associates to a natural transformation $\alpha : H_X \Rightarrow F$ the element $\bar{\alpha} = \alpha_X(1_X) \in F(X)$. Conversely, to any element $\beta \in F(X)$, $(\star)$ assigns a natural transformation $\bar{\beta} : H_X \Rightarrow F$. That is, for each $Y \in \mathcal{C}$ there is a function $\bar{\beta}_Y : H_X(Y) = \mathrm{Hom}_{\mathcal{C}}(Y, X) \to F(Y)$ defined by $\bar{\beta}_Y(f) := (F(f))(\beta) \in F(Y)$ for $f \in \mathrm{Hom}_{\mathcal{C}}(Y, X)$. For a full proof of the Yoneda lemma we refer to [Lei14].

Consequences of the Yoneda Lemma

Corollary A.1. *For any locally small category $\mathcal{C}$, the Yoneda embedding*

$$H_{\bullet} : \mathcal{C} \to [\mathcal{C}^{\mathrm{op}}, \mathbf{Set}]$$

is full and faithful.

Remark A.15. Intuitively, corollary A.1 states that for $X, Y \in \mathcal{C}$, a map $H_X \Rightarrow H_Y$ of presheaves is essentially the same thing as a map $X \to Y$ in $\mathcal{C}$.

Corollary A.2. *Let $\mathcal{C}$ be a locally small category and $X, Y \in \mathcal{C}$. Then we have*

$$H_X \cong H_Y \iff X \cong Y \iff H^X \cong H^Y.$$

Remark A.16. The above corollary states that if $\mathrm{Hom}_{\mathcal{C}}(X, Y) \cong \mathrm{Hom}_{\mathcal{C}}(X, Y')$ naturally in X, then $Y \cong Y'$. Hence, two objects are the same if and only if they 'look' the same from all viewpoints, that is if the maps into the given objects are the same.

Lemma A.6. *(Representables preserve limits). Let $\mathcal{C}$ be a locally small category and let $X \in \mathcal{C}$. Then $H^X = \mathrm{Hom}_{\mathcal{C}}(X, -) : \mathcal{C} \to \mathbf{Set}$ preserves limits.*

A.5 Adjunctions

Adjunctions play a central role throughout mathematics, capturing a certain interaction between two categories.

Definition A.50. An *adjunction* consists of a pair of functors $F : \mathcal{C} \rightleftarrows \mathcal{D} : G$ together with an isomorphism

$$\mathrm{Hom}_{\mathcal{C}}(X, G(Y)) \cong \mathrm{Hom}_{\mathcal{D}}(F(X), Y), \qquad (\star\star)$$

for any $X \in \mathcal{C}$ and $Y \in \mathcal{D}$, which is natural in X and Y, that is for any $f : X \to X'$ and $g : Y \to Y'$ we have commutative diagrams

$$\begin{array}{ccc} \mathrm{Hom}_{\mathcal{D}}(F(X), Y) & \xrightarrow{\cong} & \mathrm{Hom}_{\mathcal{C}}(X, G(Y)) \\ {\scriptstyle \mathrm{Hom}_{\mathcal{D}}(F(X), g)} \downarrow & & \downarrow {\scriptstyle \mathrm{Hom}_{\mathcal{C}}(X, G(g))} \\ \mathrm{Hom}_{\mathcal{D}}(F(X), Y') & \xrightarrow[\cong]{} & \mathrm{Hom}_{\mathcal{C}}(X, G(Y')) \end{array}$$

and

$$\begin{array}{ccc} \mathrm{Hom}_{\mathcal{D}}(F(X'), Y) & \xrightarrow{\cong} & \mathrm{Hom}_{\mathcal{C}}(X', G(Y)) \\ {\scriptstyle \mathrm{Hom}_{\mathcal{D}}(F(f), Y)} \downarrow & & \downarrow {\scriptstyle \mathrm{Hom}_{\mathcal{C}}(f, G(Y))} \\ \mathrm{Hom}_{\mathcal{D}}(F(X), Y) & \xrightarrow[\cong]{} & \mathrm{Hom}_{\mathcal{C}}(X, G(Y)) \end{array}$$

Remark A.17. Given an adjunction $F : \mathcal{C} \leftrightarrows \mathcal{D} : G$ we say that F is *left adjoint* to G and G in *right adjoint* to F. It is common to write

$$\mathcal{C} \underset{G}{\overset{F}{\underset{\longleftarrow}{\overset{\longrightarrow}{\perp}}}} \mathcal{D}$$

or simply $F \dashv G$,

Remark A.18. Given a morphism $f : F(X) \to Y$, then the natural isomorphism $(\star\star)$ sends this morphism to its *transpose* $f^\flat : X \to G(Y)$. Likewise, $(\star\star)$ associates to a morphism $g : X \to G(Y)$ its transpose $g^\sharp : F(X) \to Y$.

Example A.17. Forgetful functors between categories of algebraic structures usually have left adjoints. For instance, there is an adjunction

$U : \mathbf{Vect}_k \leftrightarrows \mathbf{Set} : F$, where U is the forgetful functor of example A.5 and F is the free functor of example A.6. The adjunction

$$\mathrm{Hom}_{\mathbf{Set}}(S, U(V)) \cong \mathrm{Hom}_{\mathbf{Vect}_k}(F(S), V)$$

says that given a set S and a vector space V, a linear map $F(S) \to V$ is essentially the same thing as a function $S \to U(V)$. Indeed, given a linear map $g : F(S) \to V$, we can define a function $\bar{g} : S \to U(V)$ by $\bar{g}(s) = g(s)$ for all $s \in S$. On the other hand, given a function $f : S \to U(V)$ we construct a linear map $\bar{f} : F(S) \to V$ by $\bar{f}(\sum_{s \in S} \lambda_s s) = \sum_{s \in S} \lambda_s f(s)$.

A.5.1 Units and Counits of Adjunctions

In this section we give an alternative description of adjunctions. To that end, take an adjunction $F \dashv G$ for

$$F : \mathcal{C} \leftrightarrows \mathcal{D} : G.$$

Naturality of the isomorphism between $\mathrm{Hom}_{\mathcal{C}}(X, G(Y))$ and $\mathrm{Hom}_{\mathcal{D}}(F(X), Y)$ means that we have a natural isomorphism of functors $\mathcal{C}^{\mathrm{op}} \times \mathcal{D} \to \mathbf{Set}$, more precisely we have

$$\mathrm{Hom}_{\mathcal{C}}(-, G(-)) \cong \mathrm{Hom}_{\mathcal{D}}(F(-), -).$$

In particular, for any object $X \in \mathcal{C}$ we obtain a morphism $X \to G(F(X))$, which is the transpose of $1_{F(X)}$. Similarly, for any object $Y \in \mathcal{D}$ we obtain a morphism $F(G(Y)) \to Y$ given as the transpose of $1_{G(Y)}$. By functoriality in X and Y we thus get natural transformations

$$\eta : 1_{\mathcal{C}} \Rightarrow G \circ F \quad \text{and} \quad \epsilon : F \circ G \Rightarrow 1_{\mathcal{D}},$$

called the *unit* η and the *counit* ϵ respectively.

Definition A.51. An *adjunction* consists of a pair of functors $F : \mathcal{C} \leftrightarrows \mathcal{D} : G$ together with natural transformations

$$\eta : 1_{\mathcal{C}} \Rightarrow G \circ F \quad \text{and} \quad \epsilon : F \circ G \Rightarrow 1_{\mathcal{D}},$$

such that the triangles

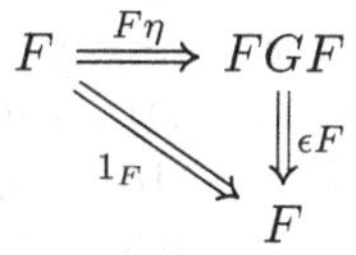

$$\begin{array}{ccc} F & \overset{F\eta}{\Longrightarrow} & FGF \\ & \searrow^{1_F} & \Downarrow \epsilon F \\ & & F \end{array}$$

and

$$\begin{array}{ccc} G & \overset{\eta G}{\Longrightarrow} & GFG \\ & \searrow^{1_G} & \Downarrow G\epsilon \\ & & G \end{array}$$

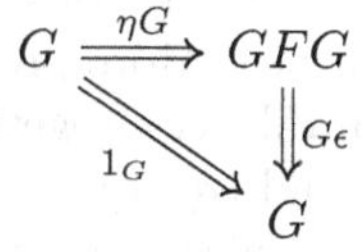

commute, i.e. the functors satisfy the *triangle identities.*

The following theorem ensures that definitions A.50 and A.51 are equivalent.

Theorem A.3. *Given a pair of functors $F : \mathcal{C} \leftrightarrows \mathcal{D} : G$. There is a one-to-one correspondence between*

- *natural isomorphisms*

$$\mathrm{Hom}_{\mathcal{C}}(X, G(Y)) \cong \mathrm{Hom}_{\mathcal{D}}(F(X), Y);$$

- *pairs of natural transformations*

$$\eta : 1_{\mathcal{C}} \Rightarrow G \circ F \quad \textit{and} \quad \epsilon : F \circ G \Rightarrow 1_{\mathcal{D}},$$

satisfying the triangle identities.

A.5.2 Adjunctions and Limits

Theorem A.4. *Let $F : \mathcal{C} \leftrightarrows \mathcal{D} : G$ be an adjunction $F \dashv G$. Then F preserves colimits and G preserves limits.*

Example A.18. Forgetful functors from categories of algebraic structures to **Set** have left adjoints, but almost never right adjoints. Hence, they preserve all limits, but rarely all colimits.

A.6 Comments on Categories with Model Structures

Many categories that we encounter in this thesis, such as the category of chain complexes $\mathbf{dgVect}_k$ or the category of differential graded algebras $\mathbf{dgAlg}_k$, come equipped with extra structure, namely with the notion of a *weak equivalence* between objects. In many situations we are interested in relations between objects only up to some form of weak equivalence. In other words, we want to formally invert weak equivalences between objects, ending up with some sort of 'homotopy category'. These ideas are encoded in categories with a *model structure*, providing an abstract setting for homotopy theoretic approaches. More precisely, a model structure on a category is a choice of weak equivalences, together with other specified classes of morphisms called *fibrations* and *cofibrations* along with some axioms that ensure that a model category provides an environment where it is technically possible to perform homotopy theory. We spell out the formal definition as given in [DS95].

Definition A.52. A *model structure* on a category $\mathcal{C}$ consists of three distinguished classes of morphisms

- *weak equivalences*;
- *fibrations*;
- *cofibrations*,

each of which is closed under composition and contains all identity morphisms. A morphism which is both a fibration (respectively cofibration) and a weak equivalence is called an *acyclic fibration* (respectively *acyclic cofibration*). We require the following axioms.

- if f and g are morphisms in $\mathcal{C}$, such that $g \circ f$ is defined, and if two of the three maps f, g, $g \circ f$ are weak equivalences, then so is the third;
- if f is a retract of g and g is a fibration, cofibration, or a weak equivalence, then so is f;
- given a commutative diagram of the following form

$$\begin{array}{ccc} A & \xrightarrow{f} & X \\ {\scriptstyle i}\downarrow & & \downarrow{\scriptstyle p} \\ B & \xrightarrow[g]{} & Y \end{array}$$

a *lift*, i.e. a morphism $h : B \to X$ such that the resulting diagram commutes, exists in either of the following two situations

  - i is a cofibration and p is an acyclic fibration;
  - i is an acyclic cofibration and p is a fibration;

- any map f can be factored in two ways
  - $f = p \circ i$, where i is a cofibration and p is an acyclic fibration;
  - $f = p \circ i$, where i is an acyclic cofibration and p is a fibration.

Definition A.53. A *model category* is a category that has a model structure and all limits and colimits.

Example A.19. The category $\mathbf{dgVect}_k^{\geq 0}$ of chain complexes, graded in non-negative degrees, can be given a model structure as follows. A morphism $f : M \to N$ is a weak equivalence if it induces an isomorphism on homology groups, so if f is a *quasi-isomorphism.* Further, the map f is a fibration, if for each $i \geq 1$ the map $f_i : M_i \to N_i$ is a surjection.

Example A.20. In the category of topological spaces **Top**, there are two natural candidates for weak equivalences, the *weak homotopy equivalences* and the *homotopy equivalences.* Both choices are part of a model structure on **Top**.

Example A.21. Another example of a model category that we encounter in this thesis is the category of simplicial sets **sSet**, which is introduced in appendix B. The weak equivalences are the *weak homotopy equivalences*, that is morphisms whose geometric realization are weak homotopy equivalences of topological spaces. The corresponding fibrations are the so called *Kan fibrations*, which are also described in appendix B.

B Simplicial Sets

Simplicial sets are essentially combinatorial models that allow to perform homotopy theory. This appendix should provide a brief, elementary introduction into the rich topic of simplicial sets. For a more extensive and nicely illustrated introduction to simplicial sets and simplicial homotopy we refer to [Fri12], which also serves as the main reference for the following.

B.1 Definitions

Categorical Definition

Definition B.1. Let $\mathbf{\Delta}$ be the category whose objects are finite ordered sets $[n] := \{0, 1, \dots, n\}$ and whose morphisms are order-preserving functions $f : [n] \to [m]$.

Definition B.2. A *simplicial set* is a functor

$$X : \mathbf{\Delta}^{\mathrm{op}} \to \mathbf{Set}.$$

Definition B.3. Let X and Y be simplicial sets. A *morphism of simplicial sets* is a natural transformation

$$f : X \Rightarrow Y.$$

Remark B.1. Simplicial sets form the category **sSet** whose objects are simplicial sets and whose morphisms are natural transformations between them.

Definition B.4. Let $\mathcal{C}$ be an arbitrary category. A *simplicial object in $\mathcal{C}$* is a functor

$$X : \mathbf{\Delta}^{\mathrm{op}} \to \mathcal{C}.$$

C. Keller, *Chern-Simons Theory and Equivariant Factorization Algebras*, BestMasters, https://doi.org/10.1007/978-3-658-25338-7

Alternative Definition

There is a natural generating set of morphisms in $\mathbf{\Delta}$. Namely, there are injections $d^i : [n] \to [n+1]$, called the *coface maps*, leaving out the i-th element in the image, and surjections $s^i : [n+1] \to [n]$, called the *codegeneracy maps*, sending two elements to the i-th element. Explicitly, the maps are defined as

$$d^i(j) = \begin{cases} j & \text{if } 0 \le j < i, \\ j+1 & \text{if } i \le j \le n \end{cases} \quad \text{and} \quad s^i(j) = \begin{cases} j & \text{if } 0 \le j \le i, \\ j-1 & \text{if } i < j \le n \end{cases}$$

for $0 \le i \le n$. One can verify that every morphism in $\mathbf{\Delta}$ can be expressed as a composite of the coface and codegeneracy maps that satisfy the following relations

$$\begin{aligned} d^j d^i &= d^i d^{j-1} \quad i < j, \\ s^j s^i &= s^i s^{j+1} \quad i \le j, \\ s^j d^i &= \begin{cases} d^i s^{j-1} & \text{if } i < j \\ 1 & \text{if } i = j, j+1 \\ d^{i-1} s^j & \text{if } i > j+1. \end{cases} \end{aligned} \tag{$\star$}$$

Example B.1. Objects in $\mathbf{\Delta}$ have a geometric realization given by the functor $\mathbf{\Delta} \to \mathbf{Top}$, defined on objects by

$$[n] \mapsto |\Delta^n|,$$

where $|\Delta^n|$ is the *standard topological n-simplex*

$$|\Delta^n| = \left\{ (t_0, \ldots, t_n) \in \mathbb{R}^{n+1} \mid t_i \ge 0, \ \sum_{i=0}^{n} t_i = 1 \right\},$$

and on morphisms $f : [n] \mapsto [m]$ by

$$f_* : |\Delta^n| \mapsto |\Delta^m|, \quad f_*(t_0, \ldots, t_n) = (s_0, \ldots, s_m), \ s_j = \sum_{f(i)=j} t_i.$$

In particular, the morphism $d^i_* : |\Delta^n| \to |\Delta^{n+1}|$ inserts a 0 in the i-th coordinate, or geometrically, it inserts $|\Delta^n|$ as the i-th face of $|\Delta^{n+1}|$. On the other hand, the morphism $s^i_* : |\Delta^{n+1}| \to |\Delta^n|$ adds the coordinates t_i and t_{i+1} and hence collapses $|\Delta^{n+1}|$ onto the standard n-simplex $|\Delta^n|$.

Let $X : \mathbf{\Delta}^{\mathrm{op}} \to \mathbf{Set}$ be a simplicial set. We write $X_n := X[n]$ and call the elements of this set *n-simplices*. Notice that in contrast to an abstract simplicial complex, the 0-simplices need not be distinct nor determine the simplex spanning them. Moreover, we write

$$d_i := Xd^i : X_n \to X_{n-1}, \quad s_i := Xs^i : X_n \to X_{n+1},$$

for all $0 \leq i \leq n$. We call the above maps *face* and *degeneracy* maps respectively. These maps satisfy relations dual to the ones given in $(\star)$. Morphisms $f : X \to Y$ between simplicial sets are thus maps of n-simplices $f_n : X_n \to Y_n$, commuting with the face and degeneracy maps. In this light, we can give an alternative definition for the notion of a simplicial set.

Definition B.5. A *simplicial set* X is a collection of sets X_n for each integer $n \geq 0$, together with *face* maps $d_i : X_n \to X_{n-1}$ and *degeneracy* maps $s_i : X_n \to X_{n+1}$ for $0 \leq i \leq n$, subjected to the following relations

$$\begin{aligned} d_i d_j &= d_{j-1} d_i \quad i < j, \\ s_i s_j &= s_{j+1} s_i \quad i \leq j, \\ d_i s_j &= \begin{cases} s_{j-1} d_i & \text{if } i < j \\ 1 & \text{if } i = j, j+1 \\ s_j d_{i-1} & \text{if } i > j+1. \end{cases} \end{aligned}$$

Remark B.2. The elements of X_n that are images of the s_i are called *degenerate*, and the rest are called *non-degenerate.*

B.2 Basic Examples

Example B.2. *(Singular simplicial set).* Let X be a topological space. Define the *singular simplicial set* of X to be the simplicial set $\mathcal{S}X$ with n-simplices defined as

$$\mathcal{S}X_n := \mathrm{Hom}_{\mathbf{Top}}(|\Delta^n|, X).$$

Let $\sigma \in \mathrm{Hom}_{\mathbf{Top}}(|\Delta^n|, X)$ be a continuous map representing a singular simplex. The face and degeneracy maps of $\mathcal{S}X$ are then given as

$$d_i \sigma = \sigma d^i_* : |\Delta^{n-1}| \to X \quad \text{and} \quad s_i \sigma = \sigma s^i_* : |\Delta^{n+1}| \to X.$$

Example B.3. *(Standard simplicial n-simplex).* Let $[n] \in \mathbf{\Delta}$ and denote by Δ^n the image of $[n]$ under the Yoneda embedding

$$\mathbf{\Delta} \to [\mathbf{\Delta}^{\mathrm{op}}, \mathbf{Set}] = \mathbf{sSet}, \quad \Delta^n := \mathbf{\Delta}(-,[n]).$$

We call the simplicial set Δ^n the *standard simplicial n-simplex* with k-simplices given by

$$\Delta^n_k = \mathbf{\Delta}([k],[n]).$$

The face and degeneracy maps are given by pre-composition in $\mathbf{\Delta}$ by d^i and s^i respectively. Explicitly we have

$$d_i : \Delta^n_k \to \Delta^n_{k-1}, \quad d_i([k] \to [n]) = [k-1] \xrightarrow{d^i} [k] \to [n],$$

$$s_i : \Delta^n_k \to \Delta^n_{k+1}, \quad s_i([k] \to [n]) = [k+1] \xrightarrow{s^i} [k] \to [n].$$

Since the Yoneda embedding is full and faithful, we have that maps of simplicial sets $\Delta^n \to \Delta^m$ are essentially the same as morphisms $f : [n] \to [m]$ in $\mathbf{\Delta}$. Concretely, a map of k-simplices is given by

$$\Delta^n_k \to \Delta^m_k, \quad ([k] \to [n]) \mapsto ([k] \to [n] \xrightarrow{f} [m]).$$

Remark B.3. The standard simplicial n-simplex Δ^n has exactly one non-degenerate element in degree n, namely $\mathrm{id}_{[n]} \in \mathrm{Hom}_{\mathbf{\Delta}}([n],[n])$. More generally, the non-degenerate k-simplices of Δ^n are the injective maps $\mathrm{Hom}_{\mathbf{\Delta}}([k],[n])$.

B.3 Simplicial Realization

We can turn simplicial sets into topological objects via the *realization functor*

$$|-| : \mathbf{sSet} \to \mathbf{Top}.$$

Definition B.6. Let X be a simplicial set. Give each set X_n the discrete topology and let $|\Delta^n|$ be the standard topological n-simplex. The *realization* $|X|$ is given by

$$|X| = \Big(\coprod_{n=0}^{\infty} X_n \times |\Delta^n| \Big) / \sim,$$

where $\sim$ is the equivalence relation generated by the relations $(x, d^i_* p) \sim (d_i x, p)$ for $x \in X_{n+1}, p \in |\Delta^n|$ and $(x, s^i_* p) \sim (s_i x, p)$ for $x \in X_{n-1}, p \in |\Delta^n|$.

Remark B.4. Intuitively, the equivalence relation $\sim$ collapses degeneracies and glues together faces. In order to see this, let us consider the above relations for an n-simplex $x \in X_n$. The first relation is that $(x, d^i_* p) \sim (d_i x, p)$, for $p \in |\Delta^{n-1}|$. Recall that d^i_* inserts $|\Delta^{n-1}|$ as the i-th face of $|\Delta^n|$ and $d_i x$ is the image of x under the i-th face map. Hence, the second term of $(x, d^i_* p)$ is a point on the i-th face of a geometric n-simplex $|\Delta^n|$. We thus interpret the first relation as taking the geometric $(n-1)$-simplex $|\Delta^{n-1}|$ that represents $d_i x$ and glue it as the i-th face in the geometric simplex $|\Delta^n|$ representing x. The second relation is $(x, s^i_* p) \sim (s_i x, p)$ for $p \in |\Delta^{n+1}|$. Recall that s^i_* collapses $|\Delta^{n+1}|$ down into $|\Delta^n|$ and $s_i x$ is a degenerate $(n+1)$-simplex. Hence, the second relation tells us that given a degenerate $(n+1)$-simplex $s_i x$ and the corresponding point $p \in |\Delta^{n+1}|$, we glue p to the n-simplex representing x at the point $s^i_* p$ by collapsing $|\Delta^{n+1}|$ onto $|\Delta^n|$.

The realization functor $|-| : \mathbf{sSet} \to \mathbf{Top}$ turns out to be adjoint to the singular set functor $\mathcal{S} : \mathbf{Top} \to \mathbf{sSet}$.

Theorem B.1. *If X is a simplicial set and Y a topological space, then*

$$\mathrm{Hom}_{\mathbf{Top}}(|X|, Y) \simeq \mathrm{Hom}_{\mathbf{sSet}}(X, \mathcal{S}Y).$$

We refer to [Fri12] for a proof.

B.4 Products of Simplicial Sets

Definition B.7. Let X and Y be simplicial sets. Their *Cartesian product* $X \times Y$ is defined by

- $(X \times Y)_n = X_n \times Y_n = \{(x, y) \mid x \in X_n, y \in Y_n\}$;
- if $(x, y) \in (X \times Y)_n$, then $d_i(x, y) = (d_i x, d_i y)$;
- if $(x, y) \in (X \times Y)_n$, then $s_i(x, y) = (s_i x, s_i y)$.

Example B.4. Let X be a simplicial set and $Y = \Delta^0$. Since Δ^0 has a unique element in each dimension we have that

$$X \times \Delta^0 \simeq X.$$

Given two simplicial sets X and Y, we want to define a simplicial set $\underline{\mathrm{Hom}}(X,Y)$ of maps from X to Y, such that we recovered the set of morphisms $\mathrm{Hom}_{\mathbf{sSet}}(X,Y)$ as the 0-simplices of this simplicial mapping space.

Definition B.8. Let $X, Y \in \mathbf{sSet}$. We define the *internal Hom* to be the simplicial set $\underline{\mathrm{Hom}}(X,Y)$ with n-simplices

$$\underline{\mathrm{Hom}}_n(X,Y) := \mathrm{Hom}_{\mathbf{sSet}}(\Delta^n \times X, Y).$$

Remark B.5. Notice that by the Yoneda lemma, there is for any simplicial set X a natural isomorphism between n-simplices of X and morphisms $\Delta^n \to X$ of simplicial sets. For the simplices of the inner Hom we thus have

$$\underline{\mathrm{Hom}}_n(X,Y) \simeq \mathrm{Hom}_{\mathbf{sSet}}(\Delta^n, \underline{\mathrm{Hom}}(X,Y)).$$

With this observation, together with definition B.8, one can check that the following adjunction relation holds between the product and the internal Hom functor in the category of simplicial sets.

$$\mathrm{Hom}_{\mathbf{sSet}}(Z, \underline{\mathrm{Hom}}(X,Y)) \simeq \mathrm{Hom}_{\mathbf{sSet}}(Z \times X, Y)$$

B.5 Simplicial Homotopy Theory

We want to use simplicial sets as combinatorial models to study homotopy theory. However, not all simplicial sets are appropriate for this intend. Similarly to the homotopy extension property for topological spaces, we need to impose an *extension condition*, known as *Kan condition*, on simplicial sets.

B.5.1 Kan Extension Condition

We first need the following definitions.

Definition B.9. A *simplicial subset* of a simplicial set X is a simplicial set Y which satisfies

$$Y_n \subset X_n$$

for all $n \geq 0$, such that the face and degeneracy maps of Y_n agree with those from X.

Definition B.10. The k-th *face* is the simplicial subset $\partial_k\Delta^n \subset \Delta^n$ generated by $d_k(\mathrm{id}_{[n]})$.

Definition B.11. The *boundary* is the simplicial subset $\partial\Delta^n \subset \Delta^n$ which is the union

$$\bigcup_j \partial_j\Delta^n$$

of all faces.

Definition B.12. The k-th *horn* is the simplicial subset $\Lambda^n_k \subset \Delta^n$, which is the union

$$\bigcup_{j \mid j \neq k} \partial_j\Delta^n$$

of all faces except the k-th one.

Remark B.6. Geometrically, the k-th horn $|\Lambda^n_k|$ on the n-simplex $|\Delta^n|$ is the subcomplex of $|\Delta^n|$ obtained by removing the interior of $|\Delta^n|$ and the k-th face.

Definition B.13. A simplicial set X satisfies the *Kan extension condition* if any morphism of simplicial sets $\Lambda^n_k \to X$ lifts to a morphism $\Delta^n \to X$ along the inclusion $\Lambda^n_k \hookrightarrow \Delta^n$.

Remark B.7. A simplicial set X that satisfies the Kan extension property is called *fibrant* or a *Kan complex.*

Example B.5. Given a topological space Y, the singular simplicial set $\mathcal{S}Y$ satisfies the Kan extension condition. Indeed, let $f : \Lambda^n_k \to \mathcal{S}Y$ be a map of simplicial sets. By the adjunction of theorem B.1, f corresponds to a continuous map $\bar{f} : |\Lambda^n_k| \to Y$. Moreover, there is a continuous map $\pi : |\Delta^n| \to |\Lambda^n_k|$ which retracts the k-th face and interior onto the remaining faces. Thus, $\bar{f}\pi : |\Delta^n| \to Y$ is a continuous map which yields the required extension $\Delta^n \to \mathcal{S}Y$.

B.5.2 Simplicial Homotopy Groups

Path Components

Throughout, let $I = \Delta^1$ and denote by $[0]$ and $[1]$ the two vertices of I.

Definition B.14. A *path* in a simplicial set X is a simplicial morphism $p : I \to X$. Equivalently, it is a 1-simplex $p \in X_1$. We call $d_1 p = p[0]$ the *initial point* of the path and $d_0 p = p[1]$ the *final point.*

Definition B.15. Two 0-simplices x and y of the simplicial set X are said to be *in the same path component* of X if there is a path with initial point x and final point y. If x and y are in the same path component, we write $x \sim y$.

The binary relation $\sim$ is in general not an equivalence relation. However, if X is fibrant, then $\sim$ is an equivalence relation [Fri12].

Theorem B.2. *If X is a Kan complex, then 'being in the same path component' is an equivalence relation.*

Remark B.8. We denote the set of path components of X by $\pi_0(X)$, that is the equivalence classes of vertices of X under the binary relation of being in the same path component.

Homotopies of Simplicial Maps

Definition B.16. Two simplicial maps $f, g : X \to Y$ are *homotopic* if there is a simplicial map $H : X \times I \to Y$ such that $H|_{X\times[0]} = g$ and $H|_{X\times[1]} = f$, that is the following diagram is commutative.

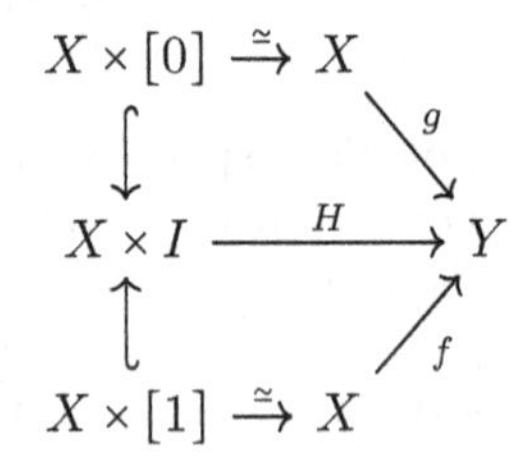

Remark B.9. We write $f \sim g$ and call f and g *homotopic* if there exists a homotopy between them.

We want that 'being homotopic' is an equivalence relation between simplicial morphisms. However, in general this is not the case, unless we work with fibrant simplicial sets, as the following theorem asserts. A proof of this result can be found in [GJ99].

Theorem B.3. *Suppose Y is a Kan complex, then homotopy of maps $X \to Y$ is an equivalence relation.*

Relative Homotopies

Definition B.17. A *simplicial map of pairs* $(X, A) \to (Y, B)$ is a simplicial map $X \to Y$ such that the image of A is contained in B.

Definition B.18. Let X be a simplicial set and A a simplicial subset of X as in definition B.9. If both X and A are Kan complexes, then (X, A) is called a *Kan pair.*

Example B.6. The *basepoint* $*$ of a simplicial set X consists of an element of X_0 and all its degeneracies. Notice that $*$ is isomorphic to Δ^0 and can be considered as the image of a simplicial mp $\Delta^0 \to X$. Since Δ^0 is a Kan complex, $(X, *)$ is a Kan pair if X is a Kan complex.

Definition B.19. Let (X, A) be a simplicial pair, where A is a simplicial subset of X, and let $f, g : X \to Y$ be two simplicial morphisms satisfying $a := f|_A = g|_A$. A *homotopy rel A* between f and g is a homotopy in the sense of definition B.16, which moreover is constant on A and hence agrees with a. In other words, we have the following commutative diagram.

$$\begin{array}{ccc} X \times I & \xrightarrow{H} & Y \\ \uparrow & & \uparrow a \\ A \times I & \twoheadrightarrow & A \end{array}$$

A proof of the following theorem can be found in [GJ99].

Theorem B.4. *Suppose Y is a Kan complex, X a simplicial set and $A \subset X$ a simplicial subset, then homotopy of maps $X \to Y$ rel A is an equivalence relation.*

Homotopy Groups

There are several equivalent ways to define homotopy groups of simplicial sets. Here, we follow the idea to define homotopy groups as homotopy classes of maps of simplicial spheres to a fibrant simplicial set as given in [Fri12].

Definition B.20. Let $(X, *)$ be a Kan pair. The *n-th homotopy group* $\pi_n(X, *)$, $n > 0$, is defined as the set of homotopy equivalence classes rel $*$ of simplicial maps $(\partial\Delta^{n+1}, *) \to (X, *)$.

Remark B.10. It is shown in [GJ99], that the simplicial homotopy groups, as well as π_0, of a simplicial set X agree with the topological homotopy groups of the geometric realization $|X|$.

B.6 Differential Forms on Simplicial Sets

We give the notion of differential forms on simplicial sets, following [Get09]. Throughout, k is a field of characteristic 0. Let Ω_n be the free commutative dg algebra over k, generated by t_i of degree 0 and dt_i of degree 1, for $0 \leq i \leq n$, quotient by the relations $T_n = 0$ and $dT_n = 0$, where $T_n := t_0 + t_1 + \cdots + t_n - 1$. More precisely, we have

$$\Omega_n := k[t_0, t_1, \ldots, t_n, dt_0, dt_1, \ldots, dt_n]/(T_n, dT_n).$$

There is a unique differential d on Ω_n such that $d(t_i) = dt_i$ and $d(dt_i) = 0$. The commutative dg algebras Ω_n are the n-simplices of a simplicial commutative dg algebra Ω defined on objects by

$$\Omega : \mathbf{\Delta}^{\mathrm{op}} \to \mathbf{cdgAlg}_k, \quad [n] \mapsto \Omega_n,$$

and on morphisms $f : [n] \to [m]$ by

$$f^* : \Omega_m \to \Omega_n, \quad t_i \mapsto \sum_{j \mid f(j)=i} t_j$$

for $0 \leq i \leq m$.

We now use Ω to define the commutative dg algebra of differential forms $\Omega(X)$ on a simplicial set X.

Definition B.21. The complex of differential forms $\Omega(X)$ on a simplicial set X is the space

$$\Omega(X) := \underline{\mathrm{Hom}}(X, \Omega)$$

of simplicial maps from X to Ω.

Remark B.11. When $k = \mathbb{R}$ is the field of real numbers, $\Omega(X)$ can be identified with the complex of differential forms on the realization $|X|$ that are polynomial on each geometric simplex of $|X|$. Hence, $\Omega(\Delta^n)$ is identified with de Rham complex of polynomial forms on the geometric n-simplex.

Printed in the United States
By Bookmasters